James Macaulay

Vivisection: Is It Scientifically Useful or Morally Justifiable?

An essay addressed specially to the medical profession

James Macaulay

Vivisection: Is It Scientifically Useful or Morally Justifiable?
An essay addressed specially to the medical profession

ISBN/EAN: 9783337254704

Printed in Europe, USA, Canada, Australia, Japan

Cover: Foto ©berggeist007 / pixelio.de

More available books at **www.hansebooks.com**

VIVISECTION:

IS IT SCIENTIFICALLY USEFUL OR MORALLY JUSTIFIABLE?

AN ESSAY

ADDRESSED SPECIALLY TO THE MEDICAL PROFESSION.

BY

JAMES MACAULAY, A.M., M.D.,

FELLOW OF THE ROYAL COLLEGE OF SURGEONS, OF EDINBURGH.

REPRINTED FROM THE LONDON EDITION.

———————

Ex iis quæ violentiâ quæruntur, alia non possunt omnino cognosci, alia possunt etiam sine scelere
—*Celsus*

———— — · —

PHILADELPHIA:

THE AMERICAN ANTI-VIVISECTION SOCIETY,

No. 1002 WALNUT STREET.

1884.

PREFACE.

THE following Essay was written in response to an offer of Two Hundred Guineas for the best Essay on "Painful Experiments on Living Animals, Scientifically and Ethically considered."

The Essays given in were adjudicated upon by seven eminent gentlemen, whose names will be found below, and who kindly consented to act as Judges. They examined the Essays submitted to them with scrupulous care, and the result was, that each of three Essays had two Judges in its favor as the best. In these circumstances, the seventh Judge did not feel warranted to decide the question, and thought it better, with the consent of all parties, to divide the award among the three authors, and publish their Essays.

NAMES OF THE JUDGES.

W. A. F. BROWNE, Esq., LL.D , F.R.C.S.E., formerly Medical Commissioner in Lunacy for Scotland, Crindau, Dumfries.

ARTHUR DE NOÉ WALKER, Esq., M.D., M.R.C.S. Eng. ; L.A.H. Member of the Royal Institution of Great Britain, 10 Ovington Gardens, London.

JAMES COWIE, Esq., M.R.C.V.S., late Vice-President, Member of Council and of the Board of Examiners of the Royal Veterinary College of Surgeons, London—Sundridge Hall, Bromley, Kent.

JOSEPH JOHNSTON, Esq., M.D., Surgeon-Major, Army Medical Department, 3 Lorne Terrace, Dublin.

JAMES GILCHRIST, Esq , M:D., Member Botanical Society, Edinburgh ; Medical Superintendent of Crichton Royal Institution, Dumfries.

DAVID JOHNSTON, Esq., M.A., M.D., L.R.C.S. Edinburgh—Kair House, Fordoun, Kincardineshire.

JOHN H. BRIDGES, Esq., M.B., F.R.C.P. London ; late Fellow of Oriel College, Oxford ; Medical Inspector Local Government Board, Whitehall, London.

VIVISECTION.

A GREAT change of opinion appears to have taken place among the medical profession in England on the subject of Vivisection. In the *Medico-Chirurgical Review* for 1842, Mr. Shaw, now (1880) the veteran Consulting Surgeon to the Middlesex Hospital, in giving a summary of his kinsman's, Sir Charles Bell's, researches, thus expressed himself:—" The profession must be well persuaded by this time what a difficult task it is to obtain any uniform results by having recourse to experiments on living animals. And it is scarcely too much to say that, if physiologists had waited patiently till cases occurred in practice, such as have actually been met with in very numerous instances, when the pathological phenomena confirmed the views deduced from anatomy, our convictions would be as strong as after all the multiplied experiments which have been performed."

The words of Sir Charles Bell himself are still more emphatic :— " In a foreign review of my former papers," he says, " the results have been considered as in favor of experimenting on living animals. They are, on the contrary, deductions from anatomy ; and I have had recourse to experiments, not to form my opinions, but to impress them upon others. It must be my apology that my utmost powers of persuasion were lost while I urged my statements on the ground of anatomy alone." And again, " Experiments have never been the means of discovery, and a survey of what has been attempted of late years will prove that the opening of living animals has done more to perpetuate error than to enforce the just views taken from anatomy and the natural sciences."—(Bell on the Nervous System.)

I have before me a printed circular, signed by thirty-eight medical men resident at Bath, which shows what was the general feeling in the profession on this subject, at a somewhat earlier period. It is dated Bath, February 27, 1825 :—" We, whose names are under-written, medical persons, chiefly practitioners, resident at Bath, do hereby engage and declare that we will, as far as in us

3

lies, prevent and discourage by our example, influence, and dissuasion, those painful and cruel anatomical experiments upon living animals, which to the disgrace of science, in our opinion, are made, sometimes without necessity or utility, and frequently without any adequate end, under the plea of promoting medical knowledge. . . . We do thus protest against and reprobate such conduct, esteeming it wholly unwarrantable and discreditable to our profession."

In the journal already quoted—the *Medico-Chirurgical Review* (vol. xxxvi), new series—at that time the leading organ of the profession, the editors, after giving some account of M. Longet's experiments, says:—" We cannot conceal our abhorrent dislike of what the French call Vivisection, in which unoffending brutes are made the victims of the most shocking sufferings, all with the view of advancing science!" More is said, in a tone of earnest indignation, with which the majority of readers in those days, no doubt, heartily sympathized.

Even so recently as 1866, when prize essays on Vivisection were published by the " Royal Society for the Prevention of Cruelty to Animals," the author of one of the essays, Dr. Markham, Physician to St. Mary's Hospital, London, while not denying the abstract right to make experiments, nor their occasional utility, observes:— " I need hardly say that courses of experimental physiology are nowhere given in this country; and that my remarks, consequently, apply only to those schools in France, and elsewhere, where such demonstrations are delivered."

In the few years that have since passed, the practice of vivisection has not only greatly increased in this country, but seems now to be regarded with altered feelings by a large part of the medical profession. There were always some physiologists and surgeons who thought it right to use this method of investigation; but their researches were quietly conducted, and not with ostentatious publicity, as in Paris and other Continental cities. The whole tone of English professional opinion was against such experiments. English medical students would have revolted against such exhibitions as were customary in foreign schools. But a change has been gradually coming over the spirit of the profession. Courses of " demonstrations in animal physiology " are given in various medical schools. "Handbooks" are published for the use of pupils in "physiological laboratories." The extent to which this mode of investigation, so recently introduced, is carried on will never be fully known; but it is already so far a department of study and education that physiological laboratories, like anatomical class-rooms, are under legal regulation and official inspection.

The tone of the medical press, which may be supposed to represent professional opinion, is also very different from what it used to be. To give but one instance—an article in the *British and Foreign Medico-Chirurgical Review*, for April, 1875, a journal holding a position analogous to that of the old *Medico-Chirurgical*, not only advocates the free practice of vivisection, but deals with objections to it as arising only from ignorance or fanaticism. The tone of the press is communicated to the profession. It may be that, at Bath, and at other centres of the Provincial Medical Association, there may be forty men ready to sign a protest as clear and firm as that which we have quoted. But the published proceedings of such bodies do not favor this hope. At the annual meeting of the British Medical Association, held at Edinburgh, in 1875, under the presidency of Sir Robert Christison, an address was delivered by Professor Rutherford, of the University of Edinburgh, who has attained eminence as an experimenter on living animals. "In recent years," said Professor Rutherford, "the teaching of physiology has made a great stride in this country. Laboratories, duly appointed, have been and are being organized. The method of physiological instruction has, in most instances, risen from the mere prelection illustrated by the diagram, to the experimental illustration of the subject. I cannot suppose that any member of this Association entertains the idea that experiments on the lower animals are not justifiable for the discovery of new truth; but I am aware that there are some who entertain the idea that vivisection is not necessary when it has for its object the mere demonstration of educational principles and facts already known. Those who hold this doctrine appear to me to forget that *physiology is an experimental science*, and that no right conception of the subject can be obtained unless the students be shown the experiments that are necessary for the demonstration of certain facts."

Professor Rutherford concluded his address by describing numerous experiments which he had made as to the effects of various medicines on the secretion of bile in the healthy dog. His description of the experiments was illustrated by diagrams, which appeared to convey to the meeting a sufficiently clear notion of his researches, and certainly in a way less disagreeable than witnessing the experiments themselves. It was a practical refutation of his own assertion that seeing the operations was essential to a right understanding of them. To these experiments I shall afterwards have to refer, but meanwhile have quoted a brief portion of the address for the sake of making a few comments.

In the first place, the Professor asserts that "physiology is an experimental science," and that, therefore, experiments on living animals must be made and must be exhibited. The fallacy in this statement is, that *experimental* is an epithet here used in a wrong sense. "Experimental science" is a synonym for "*inductive* science," or science based on the observation of facts. It is a *petitio principii* to assume that vivisection is necessary to constitute physiology one of the experimental or inductive sciences.

But passing from this, can it be said with truth that the Institutes of Medicine, or Physiology, cannot be intelligently taught without the exhibition of experiments on living animals? Professor W. P. Alison, the distinguished predecessor of Dr. Rutherford, never made such demonstrations, and he was a man as distinguished as a physician as he was successful as a teacher.

Another eminent teacher in the Edinburgh Medical School, Dr. John Fletcher, in the introductory lecture to his course on physiology and on medical jurisprudence, gave a testimony in direct opposition to that of Dr. Rutherford. "None of the functions of animals need be seen in action, in order to be perfectly well understood: they may be abundantly well fancied from preparations and representations of the organs engaged in performing them—and none, certainly, will be exhibited in action in the present lectures.

"During many years' experience in lecturing on this subject, and in delivering courses of more than ten or twelve times the duration proposed at present, I have never yet found it necessary, in a single instance, to expose a suffering animal, even to students of medicine (who are necessarily, in some degree, familiarized with sights of horror), for the purpose of elucidating any point of physiology, and I certainly shall not begin now; nor can I refrain from stating my belief, that experiments on living animals are much less necessary, even to the advancement of this science, than has been sometimes imagined. I am perfectly aware how much this plan of " interrogating Nature" has done, in modern times, for every branch of physical science; but I am equally persuaded that these advantages have been, in general, overrated—at any rate that students, in this respect, generally begin at the wrong end, and are often engaged in experimenting on animals, in hope of finding out something or other on which to found some new and surprising doctrine, while they take no manner of notice of the great number of things continually going on in their own bodies, of the rationale of which they are ignorant.

" It was a precept which I learned from my first teacher in medicine, the late venerable Abernethy, constantly to remember that I carried always about with me the best subject for observation and experiment—one the most easily to be consulted, since it was quite in my power, and one the phenomena of which should be the most interesting to me, since it was with similar beings alone that I should in future have any immediate concern; and this precept I have never lost sight of. We ought never to forget that the best subject for analysis is ourselves, and the most useful contemplation that which relates to the most common processes; and that, till we understand all which can be readily understood, with a little reflection, about ourselves, and know the rationalia of all familiar phenomena, it is preposterous to pore over the warm and quivering limbs of other animals, in search of things recondite and comparatively useless." (Introductory Lecture.)

This testimony of an experienced and successful teacher of physiology disposes of the alleged necessity for demonstrations on living animals, for purposes of explaining the facts and principles of the science. On the same point Professor Owen has recorded his opinion in these emphatic words:—"I reprobate the performance of experiments on living animals to show to students what such experiments have taught the master; whilst the arguments for learning to experiment, by repeating experiments on living animals, are as futile as those for so learning to operate chirurgically." Professor Owen thus expressed his opinion in explaining his award in the competition for the prize essay in 1866.

I have also an equally clear statement of opinion in a letter from my old teacher, Sir Robert Christison. His words are these:—"·I object to all public demonstrations by experiment on living animals, and have always done so." Sir Robert did not utter a similar protest before the British Medical Association, though he might have taken the opportunity. But there was a good deal of irritation at the time, in prospect of legislative interference ; and the profession was so far put in a defensive attitude, on account of the agitation against vivisection. The consequence was, that Professor Rutherford's address was received with apparently the unanimous approval of those present, and the report of the meeting contains no reference to any protest having been made.

It becomes an important and fitting matter for inquiry how this undoubted change in the opinion of the profession on the subject of vivisection has been brought about. How is it that so many are now advocating this new method of research, instead of keeping in the old and safe paths of observation ?

The change is certainly not due to any notable discoveries made in recent years by vivisection, nor any improvements in medical practice resulting therefrom. Still, it is interesting to inquire why the strong feeling against experimenting on living animals, which once honorably marked the profession in England, has been weakened ; and why our schools of physiology are assuming greater resemblance to the once much-censured schools of the Continent.

In attempting an explanation, I think that several things must be taken into account. In the first place, there is the natural and laudable desire for the advancement both of medical knowledge and practice. The path of progress by clinical and pathological research, though safe and sure, involves careful and patient research, as all inductive science requires. The numerous and marvelous strides made in other departments of practical science during the last half century throw into marked contrast the comparatively slow progress of medicine. To use the words of Sir James Y. Simpson :—" Ever and anon we hear it doubted, by men both without and within the profession, whether medicine has made any marked progress at all during the period that I speak of. Most of us have heard it broadly insinuated that while other departments of science and art have, during the last fifty or sixty years, been marching forwards at a pace unprecedented in their history, the art of healing has remained comparatively stationary."

To Sir James Simpson's Address " On the Modern Advancement of Physic," delivered from the Presidential chair of the Edinburgh Medico-Chirurgical Society, I shall refer presently. These sentences quoted from it express well the too prevalent feeling as to the comparatively slow progress of medicine. It is natural that both physiologists and practitioners, chafing under this feeling, hail any prospect of accelerated progress, and lend a ready ear to the assertions of those who proclaim that by vivisection they have found a shorter road to knowledge.

Another influence is to be taken into account in explaining the present attitude of the profession. The example of foreign schools in their curricula of study has been followed of late years more than it used to be in this country. " Demonstrations" in animal physiology have been introduced in various medical schools. " Physiological laboratories," and other institutions for experimental research, have been established and endowed. English as well as foreign pupils of Continental laboratories are engaged in giving practical instruction to students. Some of these professors and demonstrators are Fellows of our own Colleges of Physicians and Surgeons, and members of the Senates of our Universities. These are men of very different habits and character from some of the Continental experimenters, whose proceedings formerly caused honest indignation, but whose methods of research they are introducing. Through association and fellowship with them, there has arisen a strong *esprit de corps* in the profession, leading many to defend their brethren from attacks which have been sometimes unfair, and from charges which have been sometimes unjust. Knowing that the practice of vivisection was followed in an honest and sincere desire for the advancement of science, sympathy has been felt for the experimental physiologists, even by many who disapproved of their method of research. Expression of this sympathy by Medical Councils and Societies has given the appearance of a general feeling widely at variance with the " indignation and abhorrent dislike " expressed by the editors of the *Medico-Chirurgical Review,* in the passage already quoted, and which undoubtedly represented the opinions of the profession at that period.

A third and more marked element in the change of opinion is the discovery of anæsthetics, and their use in the performance of

A*

experiments. It is generally supposed that the use of chloroform renders impossible the horrible cruelties, especially in French laboratories, the reports of which caused Englishmen to view the whole system of vivisection with pain and dislike. The introduction of anæsthetics has thus lessened the antipathy and quieted the opposition of many professional men. But it has at the same time diverted attention from the main question, from a scientific point of view, whether vivisection is a legitimate method of research, under whatever conditions. The object of this essay is to maintain the negative, and this both on scientific and ethical grounds.

It may be well here to clear the way by a few further remarks on the use of anæsthetics in vivisection. The phrase, "painful experiments," may lead to misunderstanding. In some experiments anæsthetics are used, in others they are not used, and, in fact, would interfere with the results. In places registered under the Vivisection Act, the use of them is left to the conscience and judgment of the licensed operator. The majority of experiments, as of the public demonstrations, may be called painless; but vast numbers are not so. Many of them extend over long periods of time, during which the effect of chloroform has passed off. It is not always that the animals are "mercifully put out of pain," as one physiologist tells us is the usage at Guy's Hospital. In other London hospitals, experiments are on record where the investigations lasted day and night for weeks together. The readers of medical journals know that animals are often kept alive in a mutilated state, for the repetition or variation of experiments. Take but one instance from the *Handbook of the Physiological Laboratory* (p. 403), a demonstration upon " Recurrent sensibility:" "This can be shown only in the higher animals, the cat or dog being best adapted for the purpose The method adopted is this : The arches of one or two vertebræ are carefully sawn through, or cut through with the bone forceps, and the exposed roots very carefully freed from the connective tissue surrounding them. *If the animals be strong, and have thoroughly recovered from the chloroform, and from the operation,* irritation of the peripheral stump of the anterior root causes not only contraction in the muscles, but also movements in other parts of the body, indicative of pain. On dividing the mixed trunk the contractions cease, but the general signs of pain or sensation remain."

The Blue Book of the Royal Commission on Vivisection contains many similar facts. Dr. Klein stated, in reply to question as to use of chloroform (3605), "I prefer and use chloral hydrate; but as a general rule, for my scientific investigations, I do not use chloroform, or any other anæsthetic, except for convenience sake, in dogs and cats, and for no other animals, as a general rule." Being asked (3631) if he did not perform operations which involved a great deal of pain to the animal, the answer was: "Not as operations, but in their eventual results, we do occasionally."

Ah! it is these "eventual results" that are mainly to be considered in this question of anæsthetics. The knife may be used while the animal is under the influence of chloroform; but what of the resultant injury and mutilation, and the consequences of the experiments? Life, if not destroyed "mercifully," may be made miserable for the poor creatures. The words of Dr. George Hoggan sound strangely paradoxical at first, but they convey a sad truth nevertheless:—"*I am inclined to look upon anæsthetics as the greatest curse to vivisectable animals.* They alter too much," he explains, "the normal conditions of life to give accurate results, and they are therefore little depended upon. They, indeed, prove far more efficacious in lulling public feeling towards the vivisectors than pain in the vivisected."

So much for "painless experiments." With few and unimportant exceptions, I hold that all experiments on living animals—all, at least, to which objection is made in this essay—are painful; painful either at the time or in "eventual results," whether these be mutilation, disease or death. I hope to show that they are scientifically needless and ethically wrong, and that therefore vivisection ought to be discouraged and condemned by the medical profession.

It may be thought by some that this is an unseasonable time to renew or to extend the agitation against vivisection. An Act of Parliament, they say, has been passed, as the result of a Royal Commission of inquiry, and is now in operation, after being accepted, if not approved, by the representative bodies of the medical profession. The proceedings of experimenters are under regulations, imposed by the collective wisdom of Parliament, and with Inspectors to exercise wholesome supervision and control. Places where experiments are performed must be registered; experimenters must have licenses, either ordinary or special; and reports are made by the inspectors.   Why not wait to see how the Act works?   Such is at present the *laissez faire* tone of professional opinion, and it is largely shared by the general public.

In opposition to this spirit of indifference, I maintain that the sooner and the more fully this matter is discussed the better.   And this not in the cause of humanity only, but in the interests of science, and for the honor of our profession.

If this new system of research and of instruction is wrong, let it not have time to take deep root and to spread in our medical schools.   To the credit of the profession in Ireland, the programme of the practical course of Institutes of Medicine, under the joint direction of Trinity College, Dublin, and the College of Physicians, concludes with the significant " *N. B.—' Vivisections are absolutely prohibited.'* "   Even if this prohibition is still maintained in Ireland, we fear that it will not for some time be imitated in other schools.   The number of licensed vivisectors may vary from year to year, but the names of those to whom licenses are granted are kept secret, and the reports of the Inspectors are not open to the public.   No one can tell the nature or extent of the experimental researches, except so far as the operators choose to record them in medical journals, as Professor Rutherford has done; or to bring them before scientific societies, as Professor Ferrier has done in his Reports to the British Association.   The publication of such experiments is sure to give a fresh impulse to research by vivisection.

I think, therefore, the time has come for making an appeal to the medical profession for a calm inquiry as to the position and claims of the system.   Enough has been done to bring the matter before the general public.   There is no fear of the agitation out of doors being at an end, although the advocates of vivisection seem to think that the licensing of laboratories has silenced their opponents.   No Act of Parliament can suppress public sentiment on this question.   It will not be wise in the medical profession to set itself in direct and obstinate opposition to this public sentiment.   If, on the one side, there has been too much unintelligent clamor against vivisection, there has been, on the other side, too little of fair discussion of the merits of the case.   At the same time, a very large number of medical men have not committed themselves to the advocacy of vivisection. The great majority, I am certain, have not specially considered the subject, and have not any feeling beyond unwillingness to separate themselves from their brethren, when attacks seem to be made on the profession.

I put the question lately to the senior physician of one of our great London hospitals, if he thought vivisection had added anything to our resources which might not have been otherwise obtained, and his reply was that he had not studied the matter so as to give an answer. Another physician, occupying one of the highest positions in the profession, on my asking him about some alleged physiological discoveries, said he must inquire from his friend, P. S——, naming a surgeon and experimenter of Guy's Hospital. In the same way I have tested other medical friends, and find they are at a loss to name any practical benefits derived from vivisection. They are told that important investigations are instituted, and they are unwilling to object to any mode of research which is said to give promise of results. Comparatively few have personally studied the question; or have ventured openly to express doubt or disapproval. I believe there must be many who would be willing to see the system fairly examined, and who would even be glad to find that the result of this examination was in harmony with public sentiment, and with the former all but unanimous opinion of the medical profession in England.

It is to these men, not committed to the advocacy of vivisection, but willing to hear what can be said against as well as for it, that I address myself. It would be far more easy to write a large volume than a short essay with this purpose. In describing and analyzing the reports of physiological laboratories it would be easy to multiply proofs and illustrations of the fallacies underlying this mode of inquiry, and to point out the contradictory results of different experimenters. It would be easy, also, to gather from the records of scientific research and medical practice a great mass of observed facts and phenomena, establishing all the important conclusions which vivisection claims as discoveries. But to enter into voluminous details or minute arguments would defeat the purpose of this essay. The design of the writer is to state briefly but clearly the principles of the controversy; and by showing that vivisection is indefensible, on the ground of science as well as of sentiment, to urge medical men to re-occupy the same position which was honorably maintained by the leaders of the profession in England before this new invasion from foreign schools of physiology.

Reference has already been made to the Address by Sir James Y. Simpson, on "The Modern Advancement of Physic." It is a bright and cheering record of progress in the healing art during the first half of the present century. If any one doubts whether medicine has made marked advance, or is ever despondent as to its prospects in the future, let him read that essay, and he will find proofs of progress as great and rapid as in any department of knowledge or art. It is a retrospect at once instructive and encouraging. The enumeration of improvements both in medicine and surgery will surprise those who have not considered the state of science and of practice at an earlier period than that passed under review. Without going into many details, a few of the results may be noted.

After the middle of last century the mortality of children under five years of age, in London, was above fifty in the hundred. It is now not more than from thirty to thirty-five. The saving of life by improvement in the hygiene and management of infancy is now more than 100,000 human beings a year throughout Great Britain. The average mortality at all ages, and especially in towns, has remarkably decreased; and the chances of life have steadily increased. Some of the diseases which were formerly among the most fatal in the bills of mortality, scurvy, dysentery, ague, and smallpox, are now low in the lists. The treatment of actual disease is only one department of practical medicine. The preservation of health and the prolongation of life are equally important. These objects are attained on the large scale by the prevention of disease much more than by its cure. It may be long before specific cures are found for other fatal diseases, as effective as those which have checked the mortality from ague, scurvy, and smallpox. "But does not the history of the past," says Sir James Y. Simpson, "encourage us to a bold belief that our present most fatal diseases may, by the advancement of hygienic and medical means, be our most fatal diseases no longer?" . . I confess that I cannot but entertain an ardent belief that medical science may yet devise measures, prophylactic perhaps, rather than curative, to stay the great destruction of human life prevailing amongst us from the most fatal of these affections—phthisis. Perhaps a more advanced pathology and chemistry may yet ere long furnish us with more enlightened views of pneumonia and other inflammatory disorders than we yet possess, and arm us with more sure and potent medicinal weapons and resources against them. We have, from the experience of the last few years, every reason to hope that the whole class of zymotic diseases will be greatly subdued betimes in intensity and violence when the investigation of the physical causes predisposing to them, or even actually exciting them, is more fully expiscated.

" Besides, if by vaccination during infancy, medicine has devised means to arrest the ravages of smallpox, may it not yet devise means also, by inoculation or otherwise, to arrest the ravages of scarlet fever and measles, of hooping-cough, of typhus fever, and perhaps of the whole class of non-recurrent diseases ? And even if we fail to arrest them, we may possibly find out, for the various animal poisons producing these diseases, antidotes as certain as quinine and arsenic are antidotes against the poison of marsh fever.

" Let us at least not sit indolently down and argue ourselves into the belief that it is impossible to attain such results. The conquest of smallpox seemed to our forefathers, a hundred years ago, as impossible as the conquest of these maladies can look to any one now ; and yet we all know that the subjugation of smallpox was effected by the genius of one man, and by the devotion of one mind to its accomplishment.

" Some time before Jenner turned his attention to the subject, the learned and accomplished Dr. Mead, the first London physician of his day, wrote of the utter hopelessness of the very idea of battling with and vanquishing such a formidable enemy to human life and happiness as smallpox. He speaks of the possibility of ' a specific antidote being found against the contagions of smallpox ; ' that is, an antidote ' by which it may be so thoroughly destroyed that, though it had been received into the body, it may not produce the disease,' as an idea as wild and chimerical as that of alchemy ; and one, in his opinion, ' outraging the principles and elements of things that are so certain and well-established by the permanent laws of nature.'

" These disheartening opinions of Dr. Mead, regarding the hopelessness of ever gaining a prophylactic for smallpox, were published in 1747. Before fifty years had elapsed Jenner had both discovered and successfully applied to practice the great prophylactic measure that has rendered his name imperishable in the annals of the human race.

"Meanwhile, the prevention of diseases by the methodized avoidance of their causes has made a mighty advance during the last twenty or thirty years. Where the preceding causes of disease have been set aside from special communities by proper sanitary arrangements, human life has, in such communities, been prolonged, and the physical as well as moral health and happiness of the inhabitants have been correspondingly ameliorated."

The progress of surgery has been not less marked than that of medicine; and Sir James Y. Simpson gives a brilliant enumeration of improvements both as to operations and general treatment. In operative surgery, the abrogation of pain and suffering by anæsthetics has been a wonderful improvement; but an even greater mark of progress is the increasing endeavor to heal and to cure, without operations, cases and diseases in which operations were formerly considered indispensable. More than ever is surgery associated with medicine in the object of preservation and cure. And where operations are still required, the surgeon knows that, in eight or nine cases out of ten, the risk is not from surgical lesions, but from constitutional complications of a truly medical nature. Hence, both surgery and medicine are indebted for their progress to the better knowledge of principles which underlie every department of the healing art. "At the present day," says Sir James, "we can scarcely appreciate the vast importance of some of these branches of study, and the advantage which a knowledge of them gives us as practitioners over the cultivators of medicine half a century ago. Nor, perhaps, would it be possible to see and appreciate them in their proper value, unless we were actually again deprived of their aid—in pathology, diagnosis, and practice—and unless all the knowledge and advantages springing from them came to be suddenly obliterated and blotted out."

What, then, are the departments of research which Sir James Y. Simpson specifies as having led to the modern advancement of physic, and which give hope of future progress? The first is pathological or morbid anatomy, a branch of study, in its systematized form, almost wholly of modern growth. Secondly, pathological histology has opened up a wide field of knowledge concerning the origin, character, and courses of different diseases, and diseased actions and structures. A third department of research is that of pathological chemistry, which has afforded new and important information regarding the actual character and nature of disease. Along with these three departments of medical science—pathological anatomy, pathological histology, and pathological chemistry—the practitioner has acquired new means of physical diagnosis, by which to detect the presence or effects of morbid conditions in the living patient. The use of the microscope, and of various chemical tests to the fluid excretions of the body, has helped to improve the diagnosis of disease. Nor must we omit the improvements in materia medica, whether in the form of the remedies or in the methods of applying them, so as to exert their medicinal influence upon the body, or upon the different organs or functions of it.

On all these, and on other points, the essay of Sir James Simpson gives gratifying testimony of recent progress, with encouraging anticipations of the future. No physician in our times has been more fully acquainted with all the discoveries and researches both of English and foreign workers and authors. Yet the address contains not one word about experiments upon living animals, not one reference to those " physiological laboratories" to which many are now looking for new knowledge and power. The eye is directed throughout to the researches of legitimate science, and no help is expected from the lurid light of vivisection.

There are many, however, who have a vague idea that the beginning of all this progress was due to experiments on living animals. Let us examine the instance which is always put in the front by advocates of vivisection—the discovery of the circulation of the blood. It appears in every list of alleged discoveries from this mode of research, and would probably be the first mentioned in any controversy on the subject. When Sir William Gull was asked by the Royal Commissioners if he could " enumerate any considerable number of therapeutic remedies which have been discovered by this process of vivisection?" the answer was—" The cases bristle around us everywhere; our knowledge of dropsical affections, of pulmonary apoplexy, of enlargement of the liver, and the whole category of such affections, was due to Harvey's discovery of the circulation." Here vivisection gets credited with not only Harvey's discovery, but with

all the conseqvences of the knowledge of the circulation of the blood!
But what if this discovery was not wholly due to vivisection?

It is not necessary, in examining this question, to depreciate the
claim of Harvey to great renown, nor to inquire how far the discovery was anticipated by others, or what share they have in the
discovery. The only point here to be discussed is, " Was the discovery due to vivisection?" The Royal Commissioners say that
" Harvey *appears to have been almost entirely* indebted to vivisection
for the ever-memorable discovery of the circulation of the blood "—
the old and constant reiteration, but with more cautious assertion
than is usual.

Harvey himself did not rest his entire claims on vivisection. " I
remember," writes Robert Boyle, " that when I asked our famous
Harvey, in the only discourse I had with him (which was but a
little while before he died), what were the things that induced him
to think of a circulation of the blood, he assured me that when he
took notice that the valves in the veins of so many parts of the
body were so placed that they gave free passage to the blood towards
the heart, but opposed the passage of the blood the contrary way,
he was invited to think that so provident a cause as Nature had
not placed so many valves without a design; and no design seemed
more probable than that, since the blood could not, because of the
interposing valves, be sent by the veins to the limbs, it should be
sent through the arteries, and return through the veins, whose
valves did not oppose the course that way." It is probable that
some vivisectors do not know who Robert Boyle is, or why his
testimony is of weight, but those who do will not undervalue this
record of Harvey's own account of what led to the discovery. It
was Fabricius, of Padua, Harvey's master in anatomy, who pointed
out to him this arrangement of the valves, but Harvey's genius led
him to connect it with the various facts of the circulation already
known to Cesalpino, Servetus, and other observers. To his students
at Pisa and at Rome, Cesalpino taught the circulation from the
veins to the right side of the heart, thence to the lungs, and from
the lungs to the left side of the heart, and to the arteries. The
astonishing thing is that the complete discovery was so long delayed,
not that it came when it did. The state of science in England, far
behind that of Italy before the middle of the seventeenth century,
caused Harvey's announcements to be received with wondering

admiration. But he neither began nor completed the discovery by his experiments on living animals. He exhibited some points already known to Italian physicians, but his demonstrations failed to convince such men as Riolan, of Paris, and Hoffman, of Nuremberg. Even Dr. Willis, the biographer of Harvey, admits that " he left the doctrine of the circulation as an inference or induction only, not as a sensible demonstration. He adduced certain circumstances, and quoted various anatomical facts, which made a continuous transit of the blood from the arteries into the veins, from the veins into the arteries, a necessary consequence ; but he never saw this transit ; his idea of the way in which it was accomplished was even defective ; he had no notion of the one order of sanguiferous vessels ending by uninterrupted continuity, or by an intermediate vascular network in the other order."

It was only when Malpighi brought the microscope into play that the visible demonstration was perfect, or at least completed. What Malpighi saw in the frog's foot, Leeuwenhock saw afterwards in a tadpole, a bat's wing, and a fish's tail. When colored fluids were injected in the dead body, another form of demonstration was given.

Harvey's treatise, "De motu cordis et sanguinis circulo," beautifully systematized all that was known at his time, and his experiments demonstrate some points, but to describe the discovery as due to vivisection is an error. It is not possible to ascertain the circulation or to see it in its entirety in the living body. The very act of vivisection renders the demonstration impossible, and the discovery is due to observation of the dead body, not to experiment on the living. We shall continue to hear Harvey's name cited by vivisectors, but his own testimony is that he was first led to the discovery by anatomical observation, and by reasoning therefrom.*

Next to the circulation of the blood, the discovery of the distinct offices of the anterior and posterior roots of the spinal nerves, and the columns from which they arise, is the favorite instance of the results of vivisection. It is strange how vivisectors insist on a claim

---

* If this discovery were really of such measureless importance (or rather the part due to Harvey), we must look to the middle of the seventeenth century for the advent of a new era in the resources and the success of the practitioners of the healing art. We might expect to find from that date the death-rate wonderfully lessened, and life wonderfully prolonged. Was it so ?

whieh Sir Charles Bell has himself denied and repudiated. His express statements as to the purely anatomical source of his discovery have already been quoted. I have lately conversed on the subject with Mr. Shaw, Sir Charles Bell's friend and relative, and the able editor and expositor of his published researches. Mr. Shaw tells me that Sir Charles invariably spoke of his discovery as due to anatomical investigation ; that his experiments were performed with the utmost reluctance, and were considered by him unnecessary ; and that he often referred to the uselessness and cruelty of experiments on living animals. This is quite in accordance with the humane spirit that appears in all the writings of Sir Charles Bell.

The use of anæsthetics is also often cited as an instance of the benefit of experiments on animals. " Surely any amount of suffering that the case might have required might have been legitimately inflicted upon the lower animals, to secure such an inestimable boon to humanity." These are the words of Dr. Carpenter, a humane man as well as a distinguished physiologist, and who, when a lecturer on physiology, never exhibited experiments on living animals to his pupils. Dr. Carpenter, it will be observed, puts the case hypothetically—*might have been* legitimately inflicted. He knew that ether, and chloroform, and the anæsthetic uses of them, were not discovered by experimenting on living animals, in the sense that vivisectors wish the statement to be understood. The fact is, that the use of chloroform was the result of an experiment, and rather a perilous one, tried by Sir James Simpson upon himself, and by his assistant, Dr. Keith, as they have graphically narrated. The previous use of ether as an anæsthetic was also the result of experiments upon himself by an American dentist. Many experiments have since been performed on animals ; but the reference to anæsthetics, as an argument for vivisection, is an unworthy appeal to popular ignorance of the real state of the case.

Not less futile is the claim made as to the discovery of vaccination being due to experiments on living animals. It is well known that the discovery was made by Dr. Jenner, from observation. He observed that many of the people in the dairy district of Gloucester enjoyed a remarkable immunity from smallpox. On making inquiries he observed that cows had occasionally a pustular eruption on the udder, and that those who milked them contracted similar pustular disease on their hands. He observed that such persons enjoyed sure immunity from smallpox. He ascertained that this was the general and long-known experience of the country people. They had not reasoned on the subject, but they had observed the facts which Dr. Jenner now observed, and in consequence of which he carried on the inquiry, guided by his superior knowledge and judgment. He observed that those cows which had their udders affected had been milked by persons who had been handling horses with the affection known as "grease in the hoof." The two facts, ascertained by pure observation, were, that certain persons enjoyed immunity from smallpox, and that this immunity was due to the action on the system of another virus derived from a pustular affection in the lower animals. These observed facts really formed the basis of that discovery which has been of such incalculable benefit to the human race. The inoculation of a boy with this animal virus, instead of the smallpox matter, as then done, supplied a crucial instance and crowning test of the success of the theory. Here is Jenner's own account of this case:—

"During the investigation of the casual smallpox I was struck with the idea that it might be practicable to propagate the disease by inoculation, after the manner of the smallpox—first from the cow, and finally from one human being to another. I anxiously waited some time for an opportunity of putting this theory to the test. At length the period arrived. The first experiment was made upon a lad by the name of Phipps, in the spring of the year 1796, in whose arm a little of the vaccine virus was inserted, taken from the hand of a young woman, who had been accidentally infected by a cow. Notwithstanding the resemblance which the pustule thus excited in the boy's arm bore to variolous inoculation, yet, as the indisposition attending it was barely perceptible, I could scarcely persuade myself that the patient was secure from the smallpox. However, on his being inoculated some months afterwards, it proved that he was secure. This case inspired me with confidence; and, as soon as I could again furnish myself with virus from the cow, I made an arrangement for a series of inoculations. A number of children were inoculated in succession, one from the other; and after several months had elapsed, they were exposed to the infection of smallpox, some by inoculation, others by variolous effluvia, and some in both ways, but they all resisted it."

Let it be remarked here that the discovery was made, and the demonstration completed, so that the medical profession adopted the practice of vaccination, and the whole civilized world recognized its importance and value, before a single experiment had been made upon a living animal. A few experiments were afterwards made, not by Jenner, such as inoculating a cow with the virus from the heel of a horse; but this was not necessary to prove the efficacy of vaccination in protecting the system from smallpox. It may be said, also, that the inoculation of Phipps and the other patients were really experiments, and might have first been performed on other animals without risking human life. But experiments on lower animals, in this as in other researches, although giving some ground for reasoning by analogy, could not be accepted as conclusive. Trial of vaccination, and of subsequent exposure to smallpox infection, *must*, after any number of experiments, have been made in actual practice.

The discovery of vaccination by Jenner, and its adoption by the profession, can by no stretch of sophistry be twisted into a fair defence of vivisection. Yet we find Sir William Gull saying before the Royal Commission on Vivisection (5529, evidence):—" The whole theory of vaccination came from experiments on living animals." We cannot for a moment imagine that Sir William Gull was purposely misleading the Commissioners. The fact of a statement so unfounded being made by a man so eminent as Sir William Gull, proves how little accurate knowledge exists of the history of those discoveries on which vivisection rests its claims. Bold assertions are made, and repeated, till those not familiar with the subject receive them as true. Denials and refutations have no chance of equal attention. The public press proclaims and spreads abroad these statements, but refuses admission to counter-statements, and to arguments in disproof of the claims of vivisection.

" What has vivisection done for humanity?" This is the title of an article which appeared in the *British Medical Journal*, the organ of the British Medical Association, in January, 1875. It was at the time when there was considerable agitation, both within and beyond the profession, in consequence of the prosecution. of some medical men at Norwich, at the instance of the Society for the Prevention of Cruelty to Animals. The case attracted much public notice, and the report of the proceedings has been published in detail. It is not necessary to refer to it here, except briefly to remind my readers of the circumstances of the trial, which led, as will be seen, to events of great public importance.

At the meeting of the British Medical Association at Norwich, in 1874, M. Magnan, a French physiologist of high repute, was invited or offered to exhibit on live animals some experiments demonstrating the effects of alcohol on the system. Dogs were fastened down to the operating tables by their heads and legs, and then, through tubes inserted into their thighs, absinthe and other alcoholic fluids were injected. The operator was assisted by several medical practitioners of Norwich, and there were numerous spectators.

An eminent London surgeon was nominated as arbitrator, and allowed the experiments to continue; acting, as we are willing to believe, against his better feeling and judgment, with a desire not to seem to oppose the principle of experimenting upon living animals, rather than with direct approval of this particular series of demonstrations.

The cruel proceedings did not, however, go on without protest from some who were present. Mr. Tuffnell, President of the College of Surgeons of Dublin, loudly expressed his indignation at what he witnessed, and during one of the operations cut the tapes by which the poor victim was bound, and setting it at liberty left the place in disgust. On his way out of the house he also set free a number of cats which were shut up in a room waiting for being experimented on. The great majority remained to see the experiments.

The Royal Society for the Prevention of Cruelty to Animals very properly instituted proceedings against the Norwich medical men who assisted at the operations, M. Magnan being beyond reach of prosecution. At the trial, witnesses described the "groaning" of the dogs, their "writhing agony," and in one of them, "epileptic convulsions," all which made what was well called a "ghastly scene." Sir William Ferguson, being asked at the trial his opinion, condemned the whole exhibition as a wanton piece of cruelty. The general effects of alcohol on the system are well known; and special points, indicated by M. Magnan, could be observed in ordinary practice far more certainly than by experiments under unnatural conditions. The Norwich magistrates agreed in the opinion that the experiments were cruel and useless ; but eventually dismissed the case, as the offence did not seem to come within the meaning of the Act under which the prosecution was laid.

When the report of the trial appeared in the newspapers, and was widely circulated as a pamphlet by the Prevention of Cruelty Society, public opinion was deeply moved; and the agitation increased, till parliamentary inquiry was demanded, ending in the appointment of the Royal Commission.

The professional vivisectors and their friends naturally felt alarmed at the agitation. If other cases were brought before the English magistrates and English juries the results of the trials might be inconvenient, and would certainly be discreditable. The influence of the medical profession was therefore invoked to shelter the vivisectors from prosecution. By active efforts, both with the Government and the public press, the anti-vivisection movement was, as far as possible, countermined ; and on the appointment of the Royal Commission, two members favorable to vivisection were nominated, while scientific or medical opponents of the system were unrepresented. The influence of the General Medical Council, and of various representative bodies and eminent men in the profession, was exerted to neutralize the popular feeling against the system.

The majority of the profession acquiesced in the proceedings of the scientific defenders of vivisection, and resented the popular agitation, which was made to appear as if it were the result of ignorant and fanatical opposition to scientific research. The protest of those medical men who knew the real merits of the question was overborne. The moderate measures of Lord Hartismere and of Dr. Lyon Playfair were scouted, and the influence of the General Medical Council and the medical press, of which Mr. Erichsen and Professor Huxley were the representatives in the Royal Commission, directed the conduct of the inquiry, and led to the Report upon which the Legislature passed the present Vivisection Act.

The physiological laboratories are now protected from popular interference, and experimenters delivered from fear of prosecution under Acts previously in force. The only hope now rests in the return of the profession to the sounder scientific views which prevailed before the Continental ideas of physiological study and education found favor in England.

The Academy of Sciences at Paris, at the annual meeting, after the Norwich trial, testified its approval of M. Magnan's researches by awarding him a prize of 2500 francs. The opinion of the medical profession in England has been divided as to the acquittal of the Norwich experimenters, the majority, perhaps, approving, but a large number sharing the feeling that such experiments were not demanded in the interests of science. In order to strengthen the feeling in favor of vivisection the article in the *British Medical Journal* was prepared, to which the attention of the reader is now invited. In it we may be sure that the strongest case is put in defence of the system, and chiefly on the point of the alleged necessity of vivisection for the advancement of physiology.

The following list of discoveries is given as being due to vivisection :—

1. Discovery of the two classes of nerves, sensory and motor, by Sir Charles Bell.

2. Discovery of the functions (motor) of the seventh pair, by Sir Charles Bell. Previously to this discovery, the *portio dura* was often cut by surgeons for the cure of neuralgia.

3. Discovery of the functions of the anterior and posterior roots of the spinal nerves, by Sir Charles Bell.

4. Discovery of the functions of the anterior and posterior columns of the spinal cord, by Brown-Séquard, and others.

5. Discovery of one of the functions of the cerebellum in co-ordinating muscular movements, by Fleurens, and others.

6. Discovery of the functions of the gray matter on the surface of the cerebral hemispheres, as connected with sensation and volition, by Fleurens, Magendie, and others.

7. Discovery of the motor functions of the gray matter covering certain convolutions in the anterior part of the cerebral hemispheres, by Hitiz, Fritsch, Ferrier, Gudden and Nothnagel.

8. Demonstration of the circulation of the blood, by Harvey.

9. Measurement of the static force of the heart, and discovery of other hydraulic phenomena of the circulation, by Stephen Hales, Ludwig, etc.

10. Discovery that atmospheric air is necessary to the maintenance of life, and that when stupefied by its withdrawal, animals may be resuscitated by re-admitting it, by Robert Boyle, 1670.

B

11. Discovery that atmospheric air by continued breathing becomes vitiated and unfit for respiration, by Boyle.

12. Discovery that the air was not only vitiated but also diminished in volume by the respiration of animals, by Mayou, 1674.

13. Discovery of the relation, as regards respiration, between animal and vegetable life, by Priestley.

14. Great discoveries, by Lavoisier, on the physiology of respiration, from 1775 to 1780; namely, that oxygen is the vital element of the air, and that animals confined die when oxygen is absorbed or converted into carbonic acid, nitrogen being entirely passive.

15. Numerous facts in the physiology of digestion, observed by Blondlot, Schwann, Bernard, Lehmann and others.

16. The discovery of the functions of the lacteals, by Colin, Bernard, Ludwig, and others.

17. The discovery of the functions of the eighth pair of nerves in relation to deglutition, phonation, respiration, and cardiac action, by John Reid and others.

18. The discovery of the functions of the sympathetic system of nerves, by Pourfour de Petit, in 1727; Brachet, in 1837; John Reid, and Brown-Séquard.

19. The discovery of the phenomena of diastaltic or reflex action, by Dr. Marshall Hall.

20. The discovery of the action of light on the retina, by Homgreen, Dewar, and M'Kendrick.

21. The discovery of the glycogenic function of the liver, by Bernard, Macdonnell, and Pavy.

22. The discovery of the whole series of facts in the domain of electro-physiology, by Matteucci, Du Bois-Raymond, Pflüger, and many others.

It appears from the evidence before the Royal Commission that the article adopted as a leader in the *British Medical Journal* was prepared by Dr. J. G. M'Kendrick, Lecturer on Physiology, at Edinburgh. In reply to a question by the Commissioners (3878) as to what he thought vivisection had done for humanity, Dr. M'Kendrick referred to that published article, adding, "All of the facts which were discovered by these investigations now form, as it were, the groundwork of the knowledge of all medical men in the detection and treatment of disease." At the request of Lord Cardwell, the Chairman of the Commission, the paper in the *Medical Journal* was put in, and is reprinted in the evidence (3916).

Being asked if there had been any criticisms on the paper, Dr. M'Kendrick said—" I have not seen them myself in any journal; some one told me that he had seen a criticism or some observation about it somewhere, but I have no distinct recollection of it. I certainly did not see it " (3940).

Whether any criticism has since appeared I am not aware, but it is certainly not from the document being unanswerable, as a very brief examination of it will show.

On being asked if the list of 22 instances of the benefits derived to human beings, through the advancement of the knowledge of physiology by means of vivisection, include the whole number, the reply is: " No. I think that I have mentioned the most important which I can remember. I prepared that list with very great care at the time, and none besides occur to me at this moment."

Now, let us analyze this very carefully prepared list of discoveries alleged to be due to vivisection.

The first four refer to the discovery of the two classes of nerves, sensory and motor, and the functions of the anterior and posterior column of the spinal cord, by Sir Charles Bell and Dr. Brown-Séquard, and others.

Here is the old and reiterated assertion, as to experiment being the source of what was really due to observation. Dr. M'Kendrick, in his evidence, enforced the assertion by an illustration. If, for instance, a man was paralyzed on one side of the body, how could we tell that the paralysis was due to affection on the opposite side of the brain, without knowing that the fibres in the spinal cord cross over at the upper part of the cord to the opposite side of the brain? Lord Cardwell very shrewdly remarked that " One would have supposed that the crossing of the fibres might have been discovered by anatomy " (3879). The answer was, " The practical fact is, that it is extremely difficult, I should say almost impossible, to trace accurately the course of the fibres in the softer parts of the central nervous system."

Could there be a more unsatisfactory tone of reply? The decussation is manifest to the naked eye, and can be traced by the anatomist even in the softest parts when prepared for examination. Sir

Charles Bell himself never made this objection. The decussations have been made still more clear by sections for microscopical observation. The examination of the dead body, in cases where symptoms had been carefully observed during disease, has supplied far more useful and trustworthy facts for diagnosis and treatment than all the experiments made by physiologists on living animals.

The same remark applies to all the alleged improvements from experiments on the functions of various parts of the nervous system, including those numbered 5, 6, 7, 17, 18, in Dr. M'Kendrick's list. It is quite true that many facts and phenomena have been very conclusively shown by means of experiment, but it is not proved that observation, whether in the living or the dead body, could not afford sufficient knowledge for guidance either in the preservation of health or the treatment of disease. I maintain the sufficiency of facts obtained by observation, even in the practical uses of the alleged discovery of diastaltic or reflex action by the experiments of Dr. Marshall Hall. These experiments are constantly appealed to, especially in arguments for vivisection addressed to the profession. For medical men are quite as liable as the outside public to be led away by strong and reiterated assertion, in matters about which they have not leisure for personal and careful examination.

Dr. Marshall Hall's discovery of reflex action, it is said, has led to great improvements in the treatment of epilepsy and other nervous diseases; he discovered reflex action by experiments; therefore we must stand up for vivisection against ignorant, fanatical clamor! These are the very words with which a medical man, better known, however, as a naturalist than a practitioner, answered my inquiry as to what he thought of vivisection. It was no use trying to argue the matter with him in a passing talk. The physiologists say experiments have revolutionized medical knowledge and practice, and Marshall Hall's discovery alone is sufficient to establish their position. Is it? Let us see.

Reflex actions are those arising from the spinal cord, independent of the brain, induced by impressions on the branches of nerves, even when severed from any connection with the brain. For instance, when Mr. Bouillaud, in one of his experiments, had destroyed the cerebral lobes of a dog by red hot irons, so that there were no longer intelligent movements, still he found that the animal shrank when cold water was dashed at it, and withdrew its feet when they were pinched. Sir George Burrows gave a less repulsive example to the Royal Commissioners (186). A man in hospital is supposed to be paralyzed; the nurse tells the doctor that he must be feigning, for she saw him move his legs in the night. On being asked to move his legs, he remains motionless, and is evidently unable, though making effort of will to do so. " But if you uncover the bedclothes, and just touch the fellow's foot with a feather, he will draw his legs up, and not know that he is doing it. That is from an independent function in the spinal cord." This very simple experiment of Sir George Burrows is quite as decisive as that of M. Bouillaud, or the very horrible experiment on " recurrent sensibility " described in the Handbook for the Physiological Laboratory.

The truth is, that no experiments at all are needed for demonstrating the processes of reflex action, nor do they help towards applying the knowledge to practice, although this assertion is made. So far from leading to improved treatment of epilepsy, or other diseases supposed to be chiefly dependent on the spinal cord, the ill-digested knowledge of what Marshall Hall really did and taught has led to stupid routine, and contracted views of maladies which require most intelligent and varied treatment. This depends, in every individual case, upon conditions only to be ascertained by careful observation, or what Marshall Hall himself calls " living pathology." Apart from his experiments, no medical writer gives more shrewd and instructive remarks on the diagnosis and treatment of epileptic and other nervous diseases; but these are overlooked in the anxiety to quote his experiments in support of vivisection. It is not the multiplication of details about the nervous system that is wanted, but the wise interpretation of facts and phenomena already familiar.

So much has been said about reflex action that I have dwelt longer on the point than there was really occasion. With regard to other therapeutic or practical benefits, connected with or said to have arisen from experiments on living animals, the only one calling for distinct notice is " the abandonment of the operation of cutting the fifth pair for neuralgia." If this was often practiced one would suppose that the inefficiency of the remedy would be ascertained by

a few operations.*   But this statement is intended to convey the idea
of numbers of people remaining with distracting pain and distorted
faces, till vivisectors advised surgeons to abandon the operation!

With regard to the alleged discovery of the functions of the
several parts of the encephalon, to the experimental investigations
of which some hundreds of physiologists have devoted much labor,
there are very few results universally accepted. If we include
articles and reports in medical and scientific journals, as well as
treatises separately published, we have a huge library describing
such investigations, but the conclusions arrived at would not fill
one octavo page.   There is not a subject in the whole range of
research about which there are so many vague and so many contradictory statements.   The most recent experimenters seem to be
going over the same dreary and dismal ground as their predecessors.
Very few who are not specialists in physiological literature can
verify this assertion, which I make after comparing the reports of
contemporary vivisectors, with those of Longet, Bouillaud, Legallois,
Magendie, and Fleurens.   In fact, some of the earlier physiologists,
especially Tiedemann and Serres, can show results far more worthy
of attention than the modern vivisectors of France and Germany,
with all the superior advantages these possess in the use of anæsthetics, and in the appliances of laboratories, such as those of
Ludwig, of Leipzig; Müller, of Berlin; and Pflüger, of Bonn.
The earlier vivisectors gave due prominence to results obtained
from pathology and from comparative anatomy, and did not
maintain, like our modern physiologists of the vivisection school,
that "the whole knowledge of the animal system is derived from
experiments on living animals."

This was said in evidence repeatedly, with slight variation of
phrase, by the advocates of vivisection before the Royal Commission.   So far from such being the case, the remark of Bowman, in
his standard work on Physiology, commends itself to every unbiased
mind as true, " Vivisections upon so complex an organ as the brain
are ill-calculated to lead to useful or satisfactory results."   This is
the same conclusion at which Dr. Pritchard arrived when he said
that " the results obtained by experiments not only differ from each

---

* " *Experience has proved* that the relief, if any, is but partial and temporary,
and that the operation may, in fact, be the means of converting simple neuralgia
into irremediable structural disease."—*Miller's* "*Surgery.*"

other in essential respects, but are completely opposed to those deduced from the minute and accurate observation of pathological facts."

The next discoveries (8, 9) include the circulation of the blood, • and the various researches as to the force of the heart, the velocity of the blood, and kindred subjects. Of Harvey's discovery, and the proportion borne by vivisection in it, enough has been said. As to the experiments on the statics and dynamics of the circulation, from those of Hales to those of Ludwig, no doubt many facts have been ascertained and recorded, as is the case with all experiments, but no new or practical results appear " for the benefit of humanity." As to the absolute force of the heart considered as a hydraulic machine, and the velocity of the blood, the results of experiment vary much, and those of old Stephen Hales give probably as near an average estimate as can be expected. But for practical application in medicine the numerous experiments made since the time of Hales are quite useless. The force of the heart, for example, varies in the animals inspected, and under different conditions; and the variations are infinite in different persons, in various conditions of age, strength, and state of health. The general estimates may be interesting as facts for philosophical statement, but are useless with any view of applying such experiments to use, in maladies either of the sanguineous or nervous system. More useful information can be obtained by observing the force of the heart as indicated on the delicate dial of a balance chair, than from all the experiments of vivisectors.

From numbers 10 to 14 of Dr. M'Kendrick's list, the discoveries ascribed to vivisection need only to be named to show how futile are the claims. No painful experiments on animals were required to prove that atmospheric air is necessary for the maintenance of life; nor that atmospheric air, by continued breathing, becomes vitiated and unfit for respiration; nor that it is diminished in volume by respiration; nor to show the relation of animal and vegetable life in regard to the condition of the atmosphere. All these discoveries belong to chemistry, and were ascertained and proved by facts and occurrences in common life, and observed in ordinary course of scientific investigation. The sad tragedy of the Black Hole at Calcutta, and the frequent calamities from "choke damp" in mines, proved the effects of vitiated air, without the stupid demonstration of throwing dogs into the *grotte del cane*, far less of "experiments" by physiologists. When the interpretation of these facts was given, by the discoveries of Priestley and Lavoisier, it was a triumph of chemical, not of physiological science, and entirely apart from vivisection.

The physiology of digestion comes next (15). Numerous experiments have been made by Schwann, Bernard, and other vivisectors; but all the facts demonstrated by them, and many more, could have been ascertained by simple observation, without vivisections. If the French physiologists had taken the trouble to attend at the Parisian Abattoirs, they could have "experimented" and made observations on animals necessarily doomed to death, without injuring and destroying needless victims in their laboratories. And even in the living human subject opportunities have occurred of ascertaining all the processes of digestion, for which cruel vivisection of animals has been performed. By such experiments, in unnatural conditions of animals, no practical or useful light can be thrown on the natural processes of human digestion, in all its varieties and idiosyncracies. Abernethy, from observation and experience, knew more about digestion, and used his knowledge for the benefit of humanity, more successfully than all the vivisectors, whose practices he opposed and denounced.

With regard to the function of the lacteals, a few careful and well-directed observations, at the Abattoirs, or at any ordinary butcher's slaughter-house, would have served the same purpose as all the experiments needlessly performed in the laboratories of Colin, Bernard, and Ludwig. Anatomy had long before shown the structure and course of these vessels, and their use in regard to nutrition was well known to physiologists. In the Museum of the Royal College of Surgeons, of London, there is a remarkable series of preparations, exhibiting to anatomists the lacteals and the lymphatic vessels, and the absorbent vessels of the digestive system, injected with size, and vermilion, and mercury. No new demonstrations were needed for anatomical knowledge; and no new experiments were needed for the advancement of medical practice. Yet this is one of three notable instances which the Royal Commissioners, in their Report, describe as having been "selected for them" by Professor Turner, of Edinburgh, "in illustration of the extent to which practical medicine has been improved by physiological experiment" (p. 13).

The other two notable instances are, " the discovery of the circulation of the blood " and " Sir Charles Bell's discovery of the compound function of the spinal nerves." How far vivisection was necessary for these the reader is now prepared to judge. It is well that, in another part of their official Report, the Commissioners say, " We have not thought it part of our duty, the majority of us, not having had professional training, to decide upon matters of differing professional opinion, but we have been much struck by the consideration that severe experiments have been engaged in for the purpose of establishing results which have been considered inadequate to justify that severity, by persons of very competent authority. Cases may not improbably arise, in future, in which the physiologist may be disposed to underrate the pain inflicted in the course of establishing results which may prove to be trivial or even worthless." Dr. Samuel Johnson may be regarded as not a person of competent authority, but it is curious that he refers to this very instance of the functions of the lacteals in his celebrated paper against vivisection (*Idler*, No. 17). " I know not that by living dissections any discovery has been made, by which a single malady is more easily cured. And if the knowledge of physiology has been somewhat increased, he surely buys knowledge dear who learns the use of the lacteals at the expense of his own humanity. It is time that universal resentment should arise against those horrid operations, which tend to harden the heart and make the physician more dreadful than the gout or the stone." Dr. Johnson was, at that time, the friend and associate of the highest men in the medical profession, and would not have thus written if they approved of vivisection.

The discovery of the action of light on the retina (20) we might naturally expect to find in Dr. M'Kendrick's list, having himself contributed some experiments for its illustration. But there are few who would admit that practical knowledge on this subject depended on vivisection, any more than " the discoveries of the whole series of facts in the domain of electro-physiology." These discoveries may have " important practical bearings," but the principles on which the practice rests were the result of scientific observation, and of researches in which vivisection gave no essential aid (22).

The only remaining discovery is that of the glycogenic function

B*

of the liver (21). This has been much vaunted as an important
contribution to physiological knowledge, applicable to improvement
in medical practice. Mr. Erichsen, as spokesman of the Commissioners, made the most of it in taking the evidence of Professor
Turner. " In diabetes it was supposed, not many years ago, that
the sugar was formed in the kidneys; it is now known by physiological experiment that the sugar may be produced by a lesion of
the nervous system. Claude Bernard has shown that, if a certain
portion of the brain is injured, you get sugar in the urine; that the
sugar has nothing more to do with the kidney, and is no more a
kidney disease, in point of fact, than the purulent expectoration in
a consumptive patient has to do with the mouth; that the kidney
merely evolves it from the system, just as the mouth ejects the
purulent matter from the lungs?" To which Professor Turner
replied, "That is the case" (3126). Mr. Erichsen's question was
evidently framed for the instruction of his non-professional col-
leagues of the Commission.

The analogy suggested between the expulsion of diabetic sugar
by the kidney and of purulent sputa by the mouth was rather a
strong figure of speech; but, passing this, it was scarcely right of
Mr. Erichsen and Mr. Turner to make the Commissioners suppose
that "not many years ago sugar was believed to be formed in the
kidneys." As long ago as the time of Dr. Mead, that distinguished
physician ascribed the diabetic urine to a morbid state of the liver
and bile. A century ago Dr. Cullen taught that the morbid state
of the urine arose from the disorder of the nutritive and assimilative
functions connected with the digestive system. This was received
by the profession generally; and the melituria was understood to
indicate an abnormal result of animal chemistry, one process of
which, in natural health, was the production of sugar. What
Bernard showed was, that the formation of sugar in the liver in the
normal state is so constant that the liver may be regarded as the
sugar-producing organ. He demonstrated this by numerous
observations, especially by examining the livers of seven recently-
dead human subjects. Five of these were executed criminals. In
three healthy livers he determined the absolute weight of sugar,
finding an average of 22.03 grammes; while, in the liver of a
diabetic subject, where death was sudden, from pulmonary
apoplexy, the amount of sugar was 57.50, or more than double.

The glycogenic function of the liver was thus demonstrated in a legitimate way; and in the Abattoirs he could have performed any number of *post-mortem* experiments, if confirmation or further elucidation were desired.

But, unhappily, Bernard showed the way to experimenting on living animals. He found that, by pricking or piercing the floor of the fourth ventricle of the brain, he could increase the saccharine secretion in the liver. In this line of experimentation he has been followed by many physiologists, especially by Brunton, Pavy, Ferrier, and Schiff. In reviewing these experiments, many of which have been painful and destructive of life, I find most confused and variable results. Those results which seem the most certain are such as might be anticipated from the slightest consideration of physiological principles. Thus, it is announced that the activity of the glycogenic function is increased with an augmented flow of blood to the liver, such as takes place a few hours after a meal. On the other hand, when animals were starved—as by Dr. Brunton with rabbits, or by Dr. Wickham Legg by tying the bile ducts of cats—no irritation of the fourth ventricle will cause glycogen to appear in the liver or the urine. Schiff produced diabetes by division of the anterior columns of the spinal cord. Dr. Pavy had the same result, by dividing the superior cervical ganglion of the great sympathetic; but this lesion also caused inflammation of the lung, or pleurisy, so that the animals could not be observed for long periods, as Schiff in some cases did.

Now, all these experiments go no further than to show that the normal secretion of sugar depends on healthy action of the organs engaged in nutrition ; while unnatural interference with the actions of these organs, especially by lesion of the nervous centres by which their action is sustained, produces abnormal secretion of sugar, and diabetes. This multitude of experiments I regard as unjustifiable and needless cruelties, and leading to no useful result.

Having thus examined all the 22 alleged discoveries, claimed as due to experimental research, I must leave the reader to judge whether vivisection is "the main source of our knowledge of physiology."

A separate list is given of results "in aid of medicine." If the testimony of physicians of the highest rank in the profession is accepted, their verdict is against the alleged benefits of vivisection in the practice of the healing art. Sir James Y. Simpson did not even allude to it, in enumerating the causes of the advancement of medicine and surgery during the last half century. Professor Newman says: "I can attest that Dr. James Cowles Prichard assured me that vivisection had added nothing whatever to the physician's power of healing." When Sir Thomas Watson was giving evidence before the Royal Commission, the question was asked: "Although you have never performed any experiments, nor witnessed them, you have used the results of the experiments of others, have you not, as the basis for the advancement of your professional knowledge?" The answer was: "I have made myself acquainted with the experiments and their results, and have turned them to such uses as I could." Of this reply Mr. Macilwain, himself a distinguished and experienced surgeon, in reviewing the evidence, says: "Could any answer convey a more measured recognition of a mode of study, in reply to the question whether he had not made it a *basis* for the advancement of his professional knowledge? Could anything be more vague or unsatisfactory? Why was so experienced a witness not requested to favor the Commission with some of the details of so vast an experience? Why was he not requested to state in *what* cases he had turned it to account, and how far it had or had not answered his expectations?"

The truth is, that the question was put apparently for the sake of the lay members of the Commission, and for the non-professional readers of the Blue Book. It was intended to suggest that vivisection had been the source of improvements, if not of an entire reform of practice, in thus speaking of it as *the basis of advance in professional knowledge*. The interrogator knew too well, however, how imprudent it would be to follow up the tentative question. To have asked for details or examples would have exposed the futility of the claims of the vivisectionists to have amended or altered medical practice. Where attempts have been made to give details, the examples are not only few, but they lead at once back to the very matter under dispute, whether the knowledge on which the practice rests came from vivisection or from legitimate methods of research.

On the article in the *British Medical Journal*, already quoted, entitled "What has Vivisection done for Humanity?" the following are the examples given, under the head of benefits, in "advancing therapeutics, relief of pain," etc. :—1. Use of ether. 2. Use of chloroform. 3. Chloral discovered experimentally by Liebreich. 4. The action of all remedies are only definitely ascertained by experiments on animals. 5. Action of Calabar Bean by Fraser. 6. Antagonism between active substances and the study of antidotes.—Many observers.

Could there be a more meagre and more misleading set of examples? The practical use of anæsthetics would have been introduced and perfected if a single experiment on an inferior animal had never been made. The action of remedies on the human body can only be definitely ascertained by observation, and experiments on animals are more likely to mislead than to assist in gaining this definite knowledge. The action of some substances, such as antimony on horses and mercury on dogs, is widely different from their action on the human subject ; and the effects, both of remedies and of poisons, vary much in the different animals experimented on. Dr. Thorowgood says he has seen opium given to a pigeon, enough to kill a strong man, without any effect. Goats have been known to browse on tobacco leaves, and rabbits on belladonna, without harm. Many such anomalies have been observed, and the only certain knowledge of the influence of substances on the human subject must be obtained by observation of cases in private or in hospital practice.

In the debate on Lord Truro's "Cruelty to Animals" Bill, in the House of Lords, Earl Beauchamp adduced the fact that each year 20,000 human beings lost their lives from snake-bites, and asked if a cure for snake-bites would not be a discovery of vast importance. He intended to convey to their lordships the idea that vivisection can make this discovery. Multitudes of experiments have already been made without result. Even if an antidote should appear to have some influence on the animal operated upon, the result might be different in the human subject. The only possible way of testing alleged antidotes—and the natives of different regions profess to know this—is to apply them in actual cases of snake-bite. For such there must be frequent opportunity, if 20,000 cases occur yearly. These are experiments which can do no harm, and might lead to discovery of cure. The poisoning of animals in order to try possible remedies is a needless system of cruel experiment.

The "action of the Calabar Bean," the only distinct example specified, is no argument to adduce in such a discussion. It was reported to be a very dangerous poison, and Sir Robert Christison determined to try its effect upon himself—a very fair "experiment on a living animal;" as was that of Sir James Simpson and Dr. Keith in testing the effect of chloroform as an anæsthetic. Of course, Sir Robert Christison proceeded with extreme caution, and apportioned the dose with much care, finding the effects such as had been reported by the missionaries in Africa. He then remitted the further examination to his assistant, Dr. Fraser, who, in course of experiments, noticed the remarkable effects of the bean on the pupil. With due caution, as in Sir Robert Christison's case, this effect might have been more certainly and directly observed in the human subject, and with no more danger or inconvenience than with other poisonous substances which, in minute quantities, are used as medicines. At all events, it is trifling with the question to single out this physiological fact as an example of the improvements in medical practice due to vivisection! A stronger example would have been the action of Laburnum Bark. In a case of suspected poisoning with this substance, some trials by Christison on animals were thought necessary for the satisfaction of the jury, just as the performance of vivisections was undertaken by Sir Charles Bell for the satisfaction of the Council of the Royal Society.

Passing from poisons, in regard to which some of the most plausible apologies for experiments have been urged, other pleas put forth in the *British Medical Journal* are scarcely worthy of reply or refutation. It is said, for instance (page 56, Jan. 9, 1875), "Without vivisection-experiments, we would know almost nothing of the phenomena of inflammation." After all the observations of physicians and surgeons, of physiologists and pathologists, for successive generations, at home and abroad, we are told to look to vivisectors for almost all our knowledge of the causes, the symptoms, and the results of inflammation! The plea is preposterous, and the fact of it being seriously put forward in an article specially written in defence of vivisection is sufficient to show the groundlessness of the alleged practical benefits of this method of research.

The article concludes with the following sentences, the mere quotation of which will suffice to show the inordinate claims and pretensions of vivisection:—"To record all the facts given to physiology by experiments on animals would simply be to write the history of the science. Therapeutics is yet in its infancy; but *nearly all the facts definitely known regarding the actions of remedies have been gained by experiments on animals !* To stop experiments on animals would as surely arrest the progress of physiology, pathology and therapeutics, as an edict preventing the chemist from the use of the retort, test-tube, acids, and alkalies, would arrest the progress of chemistry." On reading this, I wondered what would be the effect of such an assertion in the minds of the intelligent readers of the *British Medical Journal*—of those, at least, outside the circles of vivisectors, and their advocates or apologists. Have all the observations of clinical medicine, of pathological anatomy, of pathological histology, and pathological chemistry been vain and fruitless? Have all the labors recorded in books of medicine and surgery, in medical reports and the transactions of societies, in practical manuals and text-books, and in our official pharmaco-pœias and dispensatories, been delusive and misleading? Almost the whole classic literature of the profession belongs to a time when, in England, the practice of vivisection was comparatively unknown, and when its results were regarded with doubt, if not with condem-nation. Have all the generalizations and conclusions of past experience been superseded by the results of this new method of research? Has the healing art, in short, been wholly revolutionized since vivisectors came into the field? The official reports of the Registrar-General, the pages of our medical journals, the case-books of our practitioners, refute the claim. Till some better statement can be given of "what vivisection has done for humanity," respectable medical men will keep to the old paths—paths of honor, and not of shame.

Assuming, for the sake of argument, that such experiments may have been of scientific value, or may have led to the discovery of scientific facts of permanent importance, could such discoveries not have been arrived at by a broad and comprehensive study of natural phenomena, or of those quasi-natural facts which are the continual accompaniments of civilization? In short, could not observation have sufficed, without experiment on living animals?

To this I give a direct answer, so far as physiology is concerned, in the words of the great Cuvier : "Nature has supplied the opportunities of learning that which experiments on the living body never could furnish. It presents us, in the different classes of animals, with nearly all possible combinations of organs, and in all proportions. There are none but have some description of organs by which they are made familiar to us; and it only is needful to examine closely the effects produced by these combinations, and the results of their partial or total absence, to deduce very probable conclusions as to the nature and use of each organ, and of each form of organ in man."

Another eminent physiologist, Dr. Carpenter, says, " Almost all our knowledge of the laws of life must be derived from observation only. Experimentation can conduct us very little farther in this inquiry. The ever-varying forms of organized beings by which we are surrounded, and the constantly-changing conditions in which they exist, present us with such numerous and different combinations of causes and effects, that it must be the fault of our mode of study, if we do not arrive at some tolerably definite conclusion as to their mutual relations." Specially, on one branch of experimental research, engaging a large share of attention in physiological laboratories, Dr. Carpenter says: " On such subjects as the functions of the different parts of the encephalon, I do not believe that experiment can give trustworthy results; such violence to one part cannot be put in practice without functional disturbance of the rest. Here I consider that a careful anatomical examination of the progressively complicated forms of the encephalon from fishes up to man—*the experiments. already prepared by nature*—is far more likely, than any number of experiments, to elucidate the problem."

No clearer statement could be given as to the value of comparative anatomy and physiology, or observation of the structure and functions of the organs of the lower animals, in the study of human physiology. I may add, that the observation of abnormal specimens of the human body is also capable of affording conclusions which experimenters seek to arrive at by their painful processes. A careful collection and arrangement of such observations would establish, and has established, many facts in physiology with far greater certainty than experiment could do. In truth, the observation of the human organs, in their early development and in cases of anomalous growth, affords many examples of " experiments prepared by nature."

Of the light thrown on physiology by the facts and laws of physical and chemical science, it is needless to speak in detail. Reference is made to these branches of science in this place, only because the advocates of experimental physiology, as we have seen in examining Dr. M'Kendrick's list of alleged discoveries, unfairly adduce facts of natural science in support of their method of research.

If physiology owes much to comparative anatomy, and also to physics, and to chemistry, it owes much to pathology. Along with pathology is included morbid anatomy, or the *post-mortem* inspection of structure, for investigation of the results of diseased action in life. When the writer was a student at the University of Edinburgh, there was a good deal of discussion about vivisection, then attracting considerable notice, from the experiments of Magendie and other French physiologists. He well remembers Dr. Abercrombie's strongly-expressed opinion about such experiments, and his advice to depend on clinical and pathological study for the knowledge that could be applied in the practice of medicine. Having had his attention thus early directed to the claims of experiment, as compared with observation, he has ever since watched the progress of vivisection; and a review of the results now, after forty years, confirms the belief that Dr. Abercrombie's opinion and advice were right. And certainly not the least injurious influence of the present rage for experimenting is its tendency to withdraw attention from seeking the advancement of physiology, as well as medicine, through clinical and pathological study.

Not professed biologists and physiologists only, but men in high position, as physicians, are echoing the strange and novel assertion, that all our most important knowledge and improved practice is derived from experiments on living animals. The experience of medical practitioners, in all the ages which are now called pre-scientific, is depreciated, and we are told to expect a new epoch in the healing art. But the more I think of it, the more I admire the courage as well as the wisdom of M. Nelaton, the distinguished surgeon, who professes to belong to the "pre-scientific" school, and declares, in opposition to the loud voice of present opinion, that there is no such thing as "scientific medicine," in the sense understood by Bernard and his admirers; and that every source of information is delusive which is not derived from direct observation of the patient.

Our own most distinguished surgeon, the late Sir William Fergusson, made an avowal nearly as emphatic. When asked if experiments had not led to the successful treatment of complaints, or the mitigation of human suffering, he replied (Vivisection Blue Book, 1049), "I may, perhaps, speak more confidently regarding surgery than any other departments in my own profession; and in surgery I am not aware of any of these experiments on the lower animals having led to the mitigation of pain, or to improvement as regards surgical details." Being asked about John Hunter's experiments, Sir William Fergusson said, that "Hunter's first experiment, if it might be so called, was done on the human subject; and it was long after he had repeated his operation on the human subject, and others had repeated it, that the fashion of tying arteries and experimenting on the lower animals originated or was developed. He had himself in early life performed such experiments, influenced by what others had done, and by the wish to come up to what they had done in such matters; but the more matured judgment of later years would not allow him to repeat what he did in earlier days. Neither was he aware that any very expert operator on the lower animals had made himself thereby an expert operator on the human subject."

Many testimonies of a similar kind could be cited from most eminent physicians and surgeons. The only reason why stronger opposition to vivisection has not been made is from the prevalence of a vague idea that benefits of a practical kind may possibly result from increased knowledge of physiological facts and phenomena. It is forgotten, meanwhile, how all the most important facts capable of being applied in practice are already set down in books on the principles of medicine and surgery.

Even in regard to pure physiology, the study of diseased action has often given the clue to the discovery of the function of organs. Physiology has learned far more from medical practice than medical practice can ever possibly gain from experiments on the lower animals. It was the study of diseases of the brain that gave the key to what knowledge we possess of the functions of the parts of the encephalon. It was by observing that paralysis of one side of the body was associated with certain diseased conditions of the opposite side of the brain, that the singular fact was established as to the right side being governed by the left side of the brain, and the left side by the right side of the brain. It was by the study of diseased conditions and their results, by observing symptoms, and by noting the pathological appearances, that the functions of the cerebral hemispheres, and of the corpus striatum, and of the optic thalamus, and other parts of the encephalon were ascertained. The wild exploration of structure and functions, under the unnatural conditions of vivisection, is more likely to retard than to expedite the knowledge of the uses and relations of the various parts of the nervous system. In our homes and our hospitals—not in physiological laboratories—we must study the human frame, in health and disease. The records of medical observation and practice contain boundless materials for induction, if the facts were carefully studied, wisely interpreted, and judiciously applied., I know of no instance where the mode of inquiry, by observation of the human system in health and disease, has retarded the dates of alleged discoveries resulting from experiments on animals. Some new discoveries will be claimed in the future as they have been in the past. But these experiments are so liable to fallacy, and in general so contradictory, that they cannot be used as guides to practice, until the facts are ascertained by scientific and professional observation.

In most cases, the experiments can have no bearing on medical practice. Professors Hitzig or Ferrier may anticipate wonderful results from connecting glycogenic function of the liver with violent injury of the brain in dogs ; but no rational practitioner would confine his treatment of diabetes to the subduing of some supposed cerebral lesion.

The discovery of antidotes to poisons is the most plausible ground on which the danger of delay in research can be pleaded. So far as this country is concerned, and in the experience of any general practitioner, ninety-nine in every hundred cases of poisoning, and even a larger proportion, are from substances with which we are perfectly familiar, and the antidotes to which are well known. Our practice in all these cases is intelligently guided by facts of physiology and of chemistry, confirmed by general experience. In very few cases, indeed, are *specific* antidotes known for poisons, and if any are proposed, their efficiency must be proved in actual practice.

Nor is it by experiments on animals that new discoveries are

likely to be made, although the claim is urged—vainly, as we have
shown—for this origin of the great "discovery" of vaccination.
If any parallel discovery is made, in regard to other fatal diseases,
it will be by "experiments" on the human body, not on animals.
Even for the benefit of animals themselves, and indirectly for the
advantage of man as having property in animals, I have great
doubt as to such experiments being ever justifiable.   The researches
of Dr. Klein, under the sanction of Mr. Simon, Medical Officer of
the Privy Council, and assisted by grants of public money, I consider wholly unjustifiable.   To produce, artificially, such distressing
diseases as typhoid fever, or pyæmia, in sheep or cattle is a bar-
barous proceeding.   So is the attempt to develope tuberculous
disease in dogs.   No practical advantage can be gained *by the artificial production of such diseases* in animals, in throwing light
either on their nature or their treatment in man.

I do not know any more striking example of the futile results of
experimental inquiry than that which was instituted some years
ago on suspended animation.   The Royal Humane Society had
received from Dr. Silvester, and other medical men, various sug-
gestions as to the best mode of treating persons apparently drowned.
The Committee referred the proposals to the Royal Medical and
Chirurgical Society, with a request for advice.   A committee of
investigation was appointed by the Royal Medical and Chirurgical
Society, consisting of the following members:—C. J. B. Williams,
M.D., F.R.S.; C. E. Brown-Séquard, M.D., F.R.S.; George Harley,
M.D.; W. S. Kirkes, M.D.; H. Hyde Salter, M.D., F.R.S.; J.
Burdon-Sanderson, M.D.; W. S. Savory, F.R.S.; and E. H. Sieve-
king, M.D.

Now, here was a clear and well-defined object of inquiry : the
purpose for which it was instituted was practical and beneficent ;
the investigators were men of science, able and experienced.
Here, if anywhere, clear and satisfactory results might be looked
for.

In pursuing the inquiry, a large number of experiments were
made upon living animals.   In the first place, the phenomena of
apnœa, in its least complicated form, were investigated—viz., when
produced by simply depriving the animal of air.   Tracheotomy
was performed upon animals fastened down to a table on their
backs, and glass tubes inserted, and secured firmly by ligature.

Through a tube thus inserted the animal could breathe freely, but the air could be at once and effectively cut off by inserting a tightly-fitting cork into the upper end of the tube. In this way a measure could be obtained of the time when respiration would cease. In order to observe in the same animals the duration of the action of the heart, long pins were inserted through the thoracic walls into some part of the ventricles. The movements of the pin indicated the motion of the heart, after the cardiac sounds had ceased to be audible. The conclusion from many experiments was that, in simple apnœa, the action of the heart continued a considerable time after the respiratory movements had ceased; a fact well known, and needing no cruel experiments to establish it.

In dogs, the average duration of the respiratory movements, after the plugging of the tube, was 4 minutes 5 seconds; the extremes being 3 minutes 30 seconds, and 4 minutes 40 seconds. The average duration of the heart's action, on the other hand, was 7 minutes 11 seconds; the extremes being 6 minutes 40 seconds, and 7 minutes 45 seconds.

Another series of experiments led to the conclusion, that a dog may be deprived of fresh supply of air during a period of 3 minutes 50 seconds, and afterwards recover without the application of artificial means, but is not likely to recover after being deprived of air for 4 minutes 10 seconds. Experiments were also made in order to measure the force of the respiratory efforts after the plugging of the glass tube.

Hitherto nothing is ascertained except that the action of the heart continues longer than that of the lungs in suspended animation, and that the death struggles in victims of suffocation vary in duration by a few seconds. On proceeding to experiments on drowning, it was found that the time of possible recovery of dogs, after immersion, was only 1 minute 30 seconds, on an average, instead of 4 minutes, from simple deprivation of air. "To what is this striking difference due?" the investigators ask. Experiments were made in order to eliminate from the inquiry the element of struggling, also the element of cold, and, lastly, the access of water to the lungs. On this latter point it was found that a dog with the windpipe plugged recovered from a longer submersion than a dog without the windpipe plugged. The conclusion from the various experiments on immersion was, that the period of death depended mainly on the entrance of water into the lungs. Violent respiratory efforts hastened this fatal result, while the action of chloroform, as diminishing such struggles, retarded death.

Experiments were next made as to the best means of resuscitation, including galvanism, venesection, cold affusion, actual cautery, and other methods ; in all the experiments the animals being suffocated in the usual way by plugging their windpipes. None of the proposed methods obtained any support from the experiments; which failed also in giving any conclusion as to the relative value of the various modes of artificial respiration.

In presenting their report to the Royal Humane Society, the Royal Medical and Chirurgical Society were able to recommend no practical suggestions as the result of their inquiry. Dr. Edward Smith gave due credit for the extent and accuracy of the facts reported, the care with which they had been ascertained, and the pains taken to estimate the influence of disturbing causes. But in reference to the practical object in the appointment of the Committee, the report, he said, failed. The Committee had not proved that any one of their inquiries was applicable to the human subject. They recorded to a second the time when various phenomena occurred in different dogs, some surviving longer than others, and some recovering more rapidly than others. But the time during which different *men* could be immersed and recover could not be proved by experiments on *dogs*, and the Committee had shown that all their plans for the restoration of drowned dogs had failed. Dr. Webster expressed regret that so much suffering had been inflicted, and the lives of so many dogs sacrificed. He hoped that in future experiments on living animals would be avoided.

On referring to the Reports of the Royal Humane Society, and making inquiry from its officers, I learn that no modifications of the method of restoring suspended animation in persons apparently drowned resulted from the experimental inquiry. Any slight modification of the method originally introduced to the Society by Dr. Silvester has arisen out of *observation on human bodies, and experience in their treatment.*

Are there not fallacies underlying such a method of interrogating Nature which, of necessity, vitiate the results ? A clearer and more forcible reply to this question could not be given than in the

words of the old Roman physician and writer on medicine, Celsus:
" It is alike unprofitable and cruel," he says, " to lay open with the
knife living bodies, so that the art which is designed for the protection and relief of suffering is made to inflict injury, and that of
the most atrocious nature. Of the things sought for by these cruel
practices, some are altogether beyond the reach of human knowledge,
and others could be ascertained without the aid of such wicked
methods of research. The appearances and conditions of the parts
of a living body thus examined must be very different from what
they are in their natural state. If, in the entire and uninjured
body, we can often, by external observation, perceive remarkable
changes, produced from fear, pain, hunger, weariness, and a thousand
other affections, how much greater must be the changes induced by
the dreadful incisions and cruel mangling of the dissector, in inter-
nal parts whose structure is far more delicate, and which are placed
in circumstances altogether unusual." These remarks of Celsus
were made in reference to the inspection of the living bodies of
human criminals, who were handed over for this purpose to the
" physiological laboratories " of the medical school of Alexandria,
and probably to other places of study. The objections to such
researches, so strongly urged by Celsus, apply with double force to
experiments on the lower animals, where the differences of function
and of structure must further diminish the chance of light being
thrown on the physiology of man in the natural condition.

That observations made by vivisection are of necessity abnormal
and liable to fallacy, reason alone might show, independently of
experience. The sources of error arise not from any contingent
cause, but from the very nature of this method of investigation.
Nature, when interrogated, reveals only what is her condition at
the moment of examination, and hence, although the permanent
and unvarying properties of inanimate matter renders the use of
experiment of paramount value, the questioning process is more
limited, and its results more uncertain, when applied to living and
sentient beings. We cannot depend on the accuracy of conclusions
respecting the normal functions of parts, if drawn from experiments
which only tell what takes place in those unnatural conditions
induced by operations. For not only are the ordinary actions of
the organs thereby often deranged or destroyed, but many causes
conspire to render still wider the difference between the observed

and the natural condition of the subjects operated upon. The deadening of pain during the actual use of the knife and other instruments, is only one element in the contrast, although chloroform itself in many cases increases the sources of fallacy and interferes with results. The excitement and terror of the animal must be taken into account; and there is abnormal action, even if the body be made insensible and unconscious. "I do not believe," says Professor Carpenter, "that on such subjects as the functions of the different parts of the encephalon, experiments can give trustworthy results; since violence to one part cannot be put in practice without functional disturbance of the rest."

Experience has confirmed these reasonable objections to experiments on living animals as necessarily liable to fallacy. The results obtained by different experimenters are so various, and often so contradictory, that there is scarcely a single position laid down by them that can with confidence be adopted. We find that the most opposite results occur at different times from injury of the same organs; that injury of different organs often produce the same results; and that the same experiments are not followed by the same results in different subjects. The latter remark applies specially to poisons, the effects of which show remarkable variations in different animals. I think that the true value of these experimental researches was rightly estimated by Dr. Pritchard, who, in his work on insanity, says :—"It is well known to all those who have paid attention to the recent progress of physiology, that attempts have been made to ascertain the functions of the different parts of the brain and its appendages by removing successively parts of these organs from living animals, and noticing the changes which ensued in their actions when thus mutilated. The most celebrated of these was the series of experiments instituted by M. Fleurens. MM. Magendie and Serres, and more lately Fodera and Bouillaud, have occupied themselves with similar researches. The results obtained from these experiments not only differ in essential respects from each other, but are completely opposed to conclusions deduced from inquiries instituted and pursued for several years on a different path. These inquirers are disposed to distrust all the results of vivisection, or experiments performed by cutting away the brains of living animals. The method of research which they have pursued is that of minute and accurate observation of pathological facts."

The following passage occurs in the work of the late Dr. Barclay, the founder of the Museum of the Royal College of Surgeons of Edinburgh, "On the Muscular Motions," p. 298 : " In making experiments on live animals, even when the species of respiration is the same as our own, anatomists must often witness phenomena that can be phenomena only of rare occurrence. After considering that the actions of the diaphragm, in ordinary cases, are different from its actions in sneezing and coughing, and these again different from its actions in laughing and hiccup; after considering that our breathing is varied by heat and cold, by pleasure and pain, by every strong mental emotion, by the different states of health and disease, by different attitudes and different exertions,—we can hardly suppose that an animal under the influence of horror; placed in a forced and unnatural attitude; its viscera exposed to the stimulus of air; its blood flowing out; many of its muscles divided by the knife; and its nervous system driven to violent desultory action from excruciating pain, would exhibit the phenomena of ordinary respiration. In that situation its muscles must produce many effects, not only of violent but irregular action ; and not only the muscles usually employed in performing the function, but also the muscles that occasionally are required to act as auxiliaries. If different anatomists, after seeing different species of animals or different individuals of the same species respiring under different experiments of torture, were each to conclude that the phenomena produced in these cases were analogous to those of ordinary respiration, their differences of opinion as to the motions of ordinary respiration would be immense."

What Dr. Barclay here says about the fallacies inseparable from experiments on respiration will apply with greater force to other departments of physiology which have been investigated in a similar manner.

It would be easy to multiply testimonies, but there is space only to add the statements of one or two experimenters who have themselves admitted the uncertain and fallacious nature of their method of research. M. Legallois remarks in one place, of his " Experiments on the influence of the nervous system on the circulation : " " J'eus presque autant de résultats différens que d' expériences ; et

c

après bien des efforts inutiles pour porter la lumière dans cette
ténébreuse question, je pris la partie de l'abandonner non sans
regret d'y avoir sacrifié un grand nombre d' animaux, et perdu
beaucoup de temps."

The experience of M. Colin, a zealous advocate and extensive
practicer of vivisection, is worthy of being noted. " Certain experiments," he says, "are complex in their nature when they are
applied to important functions, the perturbations of which react on
nearly the whole animal economy. Apply your instrument to the
brain or the heart, and immediately you have general and serious
disturbances of the system which it is necessary to disengage from
those which belong to the direct and local result of the experiment."
And again, with regard to the uncertainty of the results obtained,
M. Colin says : " Often the same experiment repeated twenty times
gives twenty different results, even when the animals are placed
apparently in the same conditions. It may even happen that the
same experiment gives contradictory results." M. Colin, after
making this admission, speaks of the necessity for multiplying
experiments : " It is not necessary to recommence in order to learn."
The fairer and more philosophical conclusion would be, with M.
Legallois, to desist from a mode of investigation which experience
has shown to be unsatisfactory, and by the very nature of it, and
of necessity, fallacious.

Sir Charles Bell said that " Vivisection has done more to perpetuate error than to enforce the just views taken from anatomy
and the natural sciences." He said this chiefly in regard to the
facts and principles of physiology. But the accusation holds good
also as to the practice of the healing art, whether in medicine or
surgery. Notable illustration of this has been given by Mr. G.
Macilwain, F.R.C.S., in a recent treatise on vivisection, being chiefly
short comments on portions of the evidence given before the Royal
Commission. He proposed to prove to the Commission that experiments on living animals were not only useless and hindering more
philosophical modes of research, but that they have been misleading,
and so productive of great practical mischief in the practice of
surgery. He was not allowed to do this, being courteously reminded
that he was not before a medical committee. But he has since
published what he intended to say, and his statements are valuable

testimonies for those who seek to know the truth on the subject. Mr. Macilwain has been very long known as an eminent and successful surgeon, and in his lectures and books he has shown himself to possess much of the shrewd insight and independent thought of his great master, Mr. Abernethy.

The two illustrations of the misleading and mischievous influence of experiments on living animals are from the practice and the writings of Sir Astley Cooper and Mr. Travers, both men most popular in their day, and whose names have still great authority in the profession. The points selected by Mr. Macilwain seem to him good illustrations of the faults which are inseparable from vivisectional inquiries.

Sir Astley Cooper thought that when the neck of the thigh bone was fractured within the capsule enclosing the hip joint, repair by bony union was impracticable, and that union if effected could only be by ligament. That this mode of union was frequent after fracture of the neck of the femur, he knew, but ligamentous union also is the mode of repair in other parts. Nay, more, it was known that sometimes surgeons, after a while, purposely allowed some degree of motion in fractured bones, where they feared that the secretion of bone might be in inconvenient excess, and where ligamentous union took place. Besides, he knew that this fracture took place most commonly in persons advanced in life, when *unusual* care is necessary as regards the utmost quiet of the limb, so that no disturbance should occur in parts which it was essential to keep in apposition, and that various circumstances rendered this, in many cases, a matter of no small difficulty. Now, all this might be said to apply, more or less, to fractures in general, but it seems to have been lost sight of or unappreciated by Sir Astley. He had got the one idea of deficient reparative power, and seems to have referred failure to nothing else. Well, to prove this, as he thought, he made some experiments on animals; and here is another feature common in vivisection. A supposition is started, contrary to or irreconcileable with many known facts, or to some obvious analogy, and then experiments are made, to see if it is true. So that, in a vast number of cases, a man commences his experiments, as Sir Astley did, with the disadvantage of a foregone conclusion. He accordingly experimented on dogs; and finding that the fractures he made in the thighs of the dogs only united by ligament, he regarded that as a confirmation of his doctrine. " Now," says Mr. Macilwain, " I will venture to affirm that not one of the circumstances necessary to the proper repair of the fractured neck of the thigh bone in the human subject could be accomplished in the dog, and especially that chief of all, the continually undisturbed condition of the injured parts. But many surgeons, both here and on the Continent, took another view of the subject, and maintained that if the parts were kept perfectly still, and so retained for the requisite time, the fractured neck of the thigh-bone would do just as well as others. Amongst these were Mr. Abernethy and Baron Larrey. Cases were successful, but were for a time met by the allegation that the fractures were outside or partially outside the capsule of the joint. As this could not be proved or disproved but by dissection, years passed during which the subject was matter of opinion. At length two or three cases occurred where opportunity was given to examine the joint after death, and the bony union of the fracture was fully established. But much evil had been done. Sir Astley was surgeon to one of the largest hospitals, and a leading teacher of surgery. Concluding that bony union could not be obtained in such cases, he recommended and adopted a practice which rendered it impossible. When the patient had been in bed a fortnight or so, and the inflammation consequent on the injury had subsided, he was made to rise and use a crutch, which, as rendering bony union impossible, necessarily involved lameness for life. The lamentable result of this practice of Sir Astley, though not warranted by a careful view of all the practical facts, but which, he concluded, his experiments on dogs seemed to confirm, can only be estimated by considering the number of cases submitted to his care, besides those of his pupils, who would probably, for a time at least, adopt the practice of their distinguished teacher.

Mr. Travers performed experiments on living dogs, causing a variety of injuries to the intestines, with a view to ascertain their powers of repair under these injuries. His inquiries seem specially to have been directed to the treatment of strangulated hernia. In writing of this disease, Mr. Travers says, that the danger of the operation in strangulated hernia is from peritonitis. That is true; but now hear the remedy he proposes. "The great means to combat this is by purgatives. If there is no peritonitis," he says, " we give purgatives to prevent it; and if there is peritonitis, we give purgatives to cure it." It must be admitted that this use of purgatives has been common in the profession; but Mr. Macilwain thinks the treatment recommended by Mr. Travers worthy of special mention, because it shows how much he erred, if not in consequence of, certainly notwithstanding, his experiments on animals. The probable explanation of the accession of peritonitis is, that where mucous and serous membranes are associated in the same organ, and the irritation of the mucous surface is accompanied by some obstacle which hinders the proper relief of the mucous, the irritation, or its effects, will be transferred to the serous membrane. To treat such a state by purgatives is an evident mistake, and sufficiently accounts for the great mortality under the treatment. Mr. Macilwain adopted successfully other measures to check inflammation, and he states that his predecessor, as surgeon to the London Truss Society—the elder Taunton—never gave purgatives, and had operated upwards of fifty times with only one or two failures. Whether Mr. Travers' treatment proceeded from what he did in his operations, or from what he neglected to do, it still illustrates the misleading character of vivisection, which failed to give useful guidance, when specially questioned by men so able and distinguished as Sir Astley Cooper and Mr. Travers.

The improvement in the mode of the ligature of arteries, introduced early in this century by Mr. Jones, has been ascribed to experiments on animals. These experiments may have confirmed his views, and satisfied others who saw them, but they were made in support of observation in the human body, which a few trials on small vessels, in the operating theatre, would have established far more speedily and surely. Yet this was presented by one witness to the Royal Commission as proving the necessity for experiments.

Nor were the arguments bearing on improvements in medical practice more conclusive. Take the experimental researches of Dr. Lauder Brunton in attempting to discover the pathology of cholera. His own account of the investigation, as given in his examination before the Commission, is as follows: " It was discovered by Moreau that by performing a certain operation upon the intestine you could get a discharge into the intestine. This discharge was discovered by Kühne to be exactly the same as was found in the intestine after cholera; so I thought, if we can find out the exact part of the nervous system that is concerned in causing this discharge, we shall probably be able to find out the part of the nervous system concerned in cholera; •and having once found that out, we may be able to get a drug that will act upon it, and thus cure cholera." There were several series of experiments, in the first of which ninety cats were used. In using the knife chloroform was given, but the animals were allowed to live some time after they recovered Dr. Brunton told the Commissioners that "they suffered a certain amount of discomfort, and *possibly*, pain, indeed, *probably*, pain, though I do not think very great pain; I think probably not much more pain than a man would suffer who had perhaps a bad attack of diarrhœa." It was also said that they were killed in four or five hours.* It is not necessary to give the details of these experiments, nor do I wish to speak of them here as cruel, as the motive of the operator may have been humane. But what are we to think of the logic that led to the inquiry, and the wisdom with which it was carried out? Moreau found he could produce a discharge into the intestine; Kühne said this discharge was the same as that found after cholera; therefore, if we can find the part of the nervous system concerned in causing this discharge, we may find some drug that will act upon it, and thus cure cholera! The prospects of a cholera specific from vivisection are not very bright. Let us hope that not many hundreds of cats or dogs may be killed, in slow torture, before the unwisdom of the inquiry is recognized. The ninety cats of "the first series" of experiments might have sufficed.

If vivisection were really the luminous and fruitful method of research which its advocates represent it to be, physiology must long ere now have been the most advanced of the sciences, and none of the mysteries of animal life would remain obscure. For the last fifty years, on the Continent, many men of high rank in science, learned and gifted, with well appointed laboratories and an unlimited supply of subjects for experiment—encouraged and applauded by the profession, and with no check or restraint from law or public opinion—have zealously cultivated this field of inquiry. In recent years many physiologists and biologists in England and America

---

* This may have been the case in some of the experiments, but in the Bartholomew Hospital Reports of Dr. Legg, there are records of other experiments where the animals lingered for weeks.

have entered into rivalry with those of France, Germany, and Italy. The experiments during the past half century may be reckoned by tens of thousands, some say even hundreds of thousands. Surely we may expect to have obtained abundant fruits from all this expenditure of labor and skill, of time and of life! Surely we may ask, what are the results of this long and unfettered investigation?

Admitting, as most medical men do, the abstract *right* to perform experiments on living animals for the advancement of science, with a view to the improvement of the healing art, the utility and value of this method of research may fairly be discussed. If unquestioned and important results could be shown, the protests against vivisection from the medical profession would be few. But vivisection has been tried and found wanting. The whole history of this branch of physiological research—from the time of Herophilus, Erasistratus, and the Egyptian operators who had living human bodies to experiment upon, down to our own day, when Professor Ferrier has to be content with the anthropoid progenitor of the human race, and Professor Rutherford with man's faithful dependent, the dog, as the subjects for examination—the whole history of vivisection, if it does not convince men of science of the entire uselessness of these modes of research, will at least force them to admit that they are of infinitely less service than it is now the custom to represent them. Take any one of the particular subjects that have most occupied the attention of experimenters—the functions of the various parts of the encephalon, for example—and what a mass of vague and absurdly discordant results appear as the fruit of all their researches! After the myriads of experiments by Legallois and Wilson Philip, Amussat and Fleurens, Magendie and Bouillaud, and by multitudes of others down to our own day, it is surely fair to ask what results can be shown. What facts are there, universally or even generally admitted, that can be truly described as the fruits of vivisection? A few conclusions, indeed, are given by experimenters as having been placed by them beyond the reach of controversy; but these few, I maintain, could have been as surely arrived at by anatomical and pathological research.

It is a matter of regret, at the same time, that most men are not satisfied by the inductions obtained by legitimate means, and require for their conviction the visible demonstrations which the vivisector offers It was thus with the great discovery of Sir Charles Bell, whose experiments, he expressly states, were performed, not for his own conviction, but for the satisfaction of others. It is the same with many of the vaunted discoveries of vivisectors, who gain ready reception and loud praise for the announcement and demonstration of facts already established by clinical and pathological research.

No opponent of vivisection denies that by means of it many facts in physiology can be demonstrated, and many phenomena of animal life illustrated. No one denies that, while it is a method of research liable to much fallacy, and often apt to mislead or even to lead to wrong conclusions, it is also capable, on some points, of giving speedy and clear demonstration of facts and phenomena. The *conclusiveness* of many experiments on living animals is not disputed. For example, M. Magendie demonstrated that cutting off the eyelids of a rabbit, and leaving bare the globe of the eye, brought on ophthalmia. MM. Bouley and Colin starved a horse, made an open wound in the throat, and injected some grains of strychnine, and the poor animal died in "characteristic convulsions." M. Fleurens removed with a knife some layers of the brain of a bird; "it immediately manifested a loss of harmony in its movements, it staggered, and fell." M. Bouillaud, who anticipated Professor Ferrier in his researches on the functions of the brain, conducted numerous experiments by injuring or removing various portions of the cerebral substance in different animals. In one of these he made an opening in the forehead of a young dog, on each side, and forced a red hot iron into the anterior lobes of the brain. "Immediately afterwards the animal, after howling violently, lay down as if to sleep. On urging it, it walked or even ran, for a considerable space; but it did not know how to avoid obstacles placed in its way, and on encountering them groaned, or even howled violently. Deprived of the knowledge of external objects, it no longer made any movements either to avoid or approach them. Yet it could still perform such motions as are called instinctive; it withdrew its feet when they were pinched, and shook itself when water was poured upon it. It turned incessantly in its cage, as if to get out, and became impatient of the restraint thus imposed." These instinctive actions continued for several days, but no improvement appeared in its intellectual power, and it was killed because its irrepressible cries disturbed the neighborhood. The anterior part of the cerebrum of another dog was removed, and the results watched for several weeks. The details are too revolting to give, but the chief result noted is that the instinctive actions were not greatly affected. It ate with voracity, and when flung into the river swam on shore and returned to the house. But it acted "like an uneducated dog, whose intellect is undeveloped. When menaced, it crouches as if to implore mercy, but does not in consequence obey. Its want of docility was remarkable; when called it did not come, but lay down and wagged its tail with an air of stupidity." Experiments very similar to those of M. Bouillaud have been conducted by Professor Ferrier at the Laboratory of the West Riding Lunatic Asylum, and elsewhere. The brain being exposed by sawing away portions of the skull, the results of injury to various parts, by knife, concentrated acids, and by electric shocks, were observed and noted. Chloroform was usually administered, but the reports of the experiments show that the animals were not continuously under its influence, and sometimes were only "partly narcotized." Here, as in the French experiments, we read of the animals exhibiting signs of pain, fear, and rage. "The animal exhibited signs of pain, screamed, and kicked out with its left hind leg, at the same time turning its head round and looking behind it in an astonished manner." "When the temporo-sphenoidal gyri were being exposed the animal bit angrily, and gnawed its own legs.* It did the same generally after irritation of the same parts." "The excitability of the brain was now well-nigh exhausted, and entirely disappeared four hours after the commencement of the experiment, during which time the exploration was kept up uninterruptedly." Or take one of Professor Rutherford's experiments upon "the biliary secretion of the dog." "Nine grains of podophylline, triturated in a mortar with some bile as a solvent, were injected into the duodenum of a dog, opened for the purpose. A rapid increase in the bile-secretion ensued; but soon it diminished, and three hours after the injection it was lower than it had ever been. In this remarkable experiment,

---

* Dr. J. Crichton Browne told the Royal Commissioners that these were "merely mechanical movements," and that the animals were unconscious of pain (3189). Dr. Ferrier said he was most careful to avoid causing pain (3228). I must refer back to what I have said about anæsthetics, pp. 17–20.

c*

therefore, the diminution of bile-secretion after podophylline was more marked than its increase; indeed, the increase might have possibly been owing to the injected bile, and not to the podophylline. Towards the close of the experiment the pulse became weak, but not excessively so. *Autopsy:* The mucous membrane of the stomach and whole length of small intestine were intensely red. The small intestine contained a large quantity of fluid. The large intestine contained a considerable quantity of liquid fecal matter. There was, therefore, abundant evidence that excessive purgation was imminent." The conclusion was that this large dose of podophylline, with a biliary solvent, produced intense irritation of the intestine, with signs of purgative power, but with effect on the liver not corresponding to the other results. Other experiments showed that when the intestinal irritation is less the biliary secretion is larger. The practical use of the experiments we are not now considering, but in these, as in other operations of the vivisectors, certain physiological facts are clearly exhibited. The same remark may be made as to the vast majority of experiments given in the "Handbook of the Physiological Laboratory," and in other manuals of vivisection. Our contention is not that such experiments are *inconclusive* in many instances, but that they are *useless,* and therefore cruel and immoral. Of many of them we affirm that the facts ascertained have no bearing either on the general principle of physiology, nor on the practice of medicine. And of others, which seem to bear some relation to the advancement of knowledge or art, we affirm that the results could be and are attained by clinical and pathological observation. In the words of Celsus, which may be taken as the motto for this essay, " Hæc cognoscere prudentem medicum non cædem sed sanitatem molientem ; idque per misericordiam discere, quod alii dirâ crudelitate cognoverint." And again, " Ex iis quæ violentiâ quæruntur, alia non possunt omnino cognosci, alia possunt etiam sine scelere."

We are now prepared for considering the question, Are experiments on living animals morally justifiable? The question cannot receive a direct and categorical reply, irrespective of motives and of results. Man's dominion over the lower animals is very large, and it is his, not only by superior knowledge and power, but by Divine appointment. The dominion is not absolute, but limited by the eternal obligations of justice and mercy. Man may use this delegated dominion for his own benefit, but he may not abuse it. The gentle and genial poet, Cowper, has well expressed the extent and the limit of this dominion :—

> " The sum is this,—if man's convenience, health,
>   Or safety interfere, his rights and claims
>   Are paramount, and must extinguish theirs :
>   Else they are all, the meanest things that are,
>   As free to live, and to enjoy that life,
>   As God was free to form them at the first,
>   Who in His sovereign wisdom made them all."

For the food, the clothing, and other uses of man, many animals are sacrificed. None but Brahmins, on religious grounds, or some Vegetarians, on the grounds partly dietetic and partly ethical, object to taking the life of the lower animals for such purposes. Even when life is not taken, animals may be put to pain and may suffer injury, as in castration, to fit them for the useful service of man. On the same principle, it can fairly be argued that man has right to use animals for researches that may lead to the restoration or the preservation of human health. But, before we admit this, we must be satisfied that the results of these researches are such as justify the resort to them, and also that these results can be obtained in no other way. This is what we have investigated in the previous part of our essay, and have concluded that, on this plea, they are not justifiable.

There is another way of looking at the question of vivisection, as tending to human benefit. If it is right to perform experiments on living bodies for advancement of the healing art, why not perform them on human bodies? It has been done in past times, and may be proposed again. If condemned malefactors were operated upon, it would only be anticipating, by a brief period, their hour of death. Or the experiments might be made on the insane and imbecile, or persons defective in intellectual or moral faculties, but with animal life in natural vigor. These subjects would be free from the objections arising out of the different structure, constitution, and functions of the lower animals, though still liable to certain fallacies inseparable from the very method of research. Vivisectors make light of these alleged fallacies, and think their experiments full of light and fruit. Fair argument might be used for experiments on living men, with or without anæsthetics, as the inquiry might demand. It might be argued that it is expedient or right that one or a few should suffer for the benefit of the human family. And if the argument, "*in majus bonum*," were strengthened by reference to *corpora vilia*, then of malefactors doomed to die, and of imbeciles, this could be truly said.

Vivisectors would hardly venture, at least not yet—at least in England—to propose returning to the practice of experiments on living human bodies. Public opinion, and medical opinion, would revolt from the proposal, if biologists and physiologists should propose it.

And why? Not because the arguments for such experiments are weak, but because the objections of the moral sense are strong. Except for self-defence or self-preservation, the moral sense recoils from the infliction of pain and injury, even when a lofty motive may be urged. Why has trial by torture been banished from the jurisprudence of every civilized nation? The object of the rack, and the thumbscrew, and of all the infernal apparatus in use in our courts of law at no very remote period, was not to cause pain, far less to give any satisfaction or pleasure. The discovery of truth was the object in this method of interrogation; and with this end in view, the use of torture was justified, and directed by rulers and judges in other respects humane as well as just. In the still more horrible tortures of the Inquisition, the object was not avowedly that of vindictive punishment; nor need we assume that even the lowest executioners and officers of that dark tribunal took pleasure in the agonies of their heretic victims. The professed aim was higher even than in the processes of ordinary torture in courts of law. The advancement of Divine truth and of sacred science, or theology, was the alleged design of the Inquisition, while the spiritual welfare and eternal salvation of men might be also attained, through subjecting them to short though sharp affliction. Yet examination by torture is advocated by no one, because the infliction of pain, even for the advancement of truth, is not justifiable. And does not this apply with equal force to experiments on the lower animals? This is, indeed, interrogating nature by torture! You might operate on human subjects with no higher intelligence, and of no higher moral condition, and certainly with no more sensitive frame, than the poor brutes that are carried to the vivisector's laboratory. It is the infliction of pain and injury that cannot be justified, whether the victim be an imbecile human idiot, or a docile intelligent dog.

Professor Newman, in a published letter, has said, " Evidently the reason why it is wicked to torture a man is not because he has an immortal soul, but because he has a highly sensitive body; and so has every vertebrate animal, especially the warm-blooded. If we have no moral right to torture a man, neither have we a moral right to torture a dog." And again, " We have to add to our morals a new chapter on the Rights of Animals. Men who teach to trample them down are teachers of hard-heartedness, and are real enemies of mankind, while they undertake to promote human welfare."

There is a remarkable passage in the works of Jeremy Bentham, applying the principle of natural law to the rights of animals. It is quoted by Sir Arthur Helps in his " Talks about Animals and their Masters." " The day may come when the rest of the animal creation may acquire those rights which never could have been withheld from them but by the hand of tyranny. It may come one day to be recognized that the number of legs, the villosity of the skin, or the termination of the *os sacrum*, are reasons insufficient for abandoning a sensitive being to the caprice of a tormentor. What else is it that should trace the inseparable line? Is it the faculty of reason, or perhaps the faculty of discourse? But a full-grown horse or dog is beyond comparison a more rational as well as a more conversable animal than an infant of a day, or a week, or even a month old. But suppose the case were otherwise, what could it avail? The question is not, ' Can they reason?' nor ' Can they speak?' but 'Can they suffer?'"

If Justice requires that the rights of animals should be respected, and questions of wrong-doing not be confined to man's treatment of his fellow-men, much more does Mercy refuse to recognize the arbitrary limit of our own species. " There is implanted by Nature," says Lord Bacon, " in the heart of man, a noble and excellent affection of mercy, extending even to the brute animals which, by the Divine appointment, are subjected to his dominion."

Dr. Chalmers, in his eloquent sermon, says of humanity to the lower animals:—" It is a virtue which oversteps, as it were, the limits of a species, and which prompts a descending movement on our part, of righteousness and mercy towards those who have an inferior place to ourselves in the scale of creation.   It is not the circulation of benevolence within the limits of one species.   It is the transmission of it from one species to another.   The first is the charity of a world.   The second is the charity of a universe.   Had there been no such charity, no descending current of love and of compassion from species to species, what, I ask, would have become of ourselves? . . .   The distance upward between us and that mysterious Being who let Himself down from Heaven's high concave upon our lowly platform, surpasses by infinity the distance downward between us and everything that breathes.   And He bowed Himself thus far for the purpose of an example, as well as for the purpose of an expiation, that every Christian might extend his compassionate regards over the whole of sentient and suffering nature."   By Dr. Chalmers the duty of mercy to animals was thus lifted to the highest level of Christian ethics.   In the same spirit are the words of a distinguished man of science and Christian philanthropist, Dr. George Wilson :—"There is an example as well as a lesson for us in the Saviour's compassion for men.   Inasmuch as we partake with the lower animals of bodies exquisitely sensitive to pain, and often agonized by it, we should be slow to torture creatures who, though not sharers of our joys, or participators in our mental agonies, can equal us in bodily suffering.   We stand, by Divine appointment, between God and His irresponsible subjects, and are as gods to them."

May we not say that vivisection is thus contrary alike to the justice which regards the rights of animals, and to the mercy which has sympathy with the helpless and the suffering?   In the principle of the thing, man has no more right to perform painful or injurious experiments on animals than on human beings.

I have said that man's dominion over all living creatures is not absolute, but limited by the eternal obligations of justice and mercy. It is also to be regarded not merely as a right, but as a trust.   On this point I quote some sentences from a remarkable speech by the great Lord Erskine, when he was trying to induce the Government of his day to legislate for the protection of animals from cruelty :—
" That the dominion of man over the lower world is a moral trust, is a proposition which no man living can deny, without denying

the whole foundation of our duties. If, in the examination of the qualities, powers, and instincts of animals, we could discover nothing else but their admirable and wonderful construction for man's assistance; if we found no organs in the animals for their own gratification and happiness—no sensibility to pain or pleasure—no senses analogous, though inferior, to our own—no grateful sense of kindness, nor suffering from neglect or injury; if we discovered, in short, nothing but mere animated matter, obviously and exclusively subservient to human purposes, it would be difficult to maintain that the dominion over them was a trust, in any other sense, at least, than to make the best use for ourselves of the property which Providence had given us. But it calls for no deep or extended skill in natural history to know that the very reverse of this is the case, and that God is the benevolent and impartial author of all that He has created. For every animal which comes in contact with man, and whose powers and qualities and instincts are obviously adapted to his use, Nature has taken care to provide, and as carefully and bountifully as for man himself, organs and feelings for its own enjoyment and happiness." "The animals are given for our use, but not for our abuse. Their freedom and enjoyments, when they cease to be consistent with our just dominion and enjoyments, can be no part of their natural rights; but whilst they are consistent, their rights, subservient as they are, ought to be as sacred as our own."

Having stated the ethical principles on which the opposition to vivisection is founded, and shown that the system is not in harmony with the moral government of the world, there remains an important practical question as to the moral effects of this mode of research. Is not the tendency to harden the operator, and blunt his moral sense? And, if so, is not the system injurious, not only to those engaged in it, but to the tone and character of the medical profession, and to society at large?

In examining the question, in its moral and social bearings, it is of no avail to say that some vivisectionists are good and exemplary, and even tender-hearted men. This is true; and it may be also admitted that, in the performance of experiments, they themselves are subjected to much mental distress. Nothing but a high sense of duty, and an earnest desire to obtain useful results, could induce the medical men of culture and ordinary feeling to engage in some of the researches, the mere descriptions of which cannot be read without pain and horror. Professor Rolleston, of Oxford, in giving his evidence before the Royal Commission, bore testimony, from personal intercourse and friendship, as to the amiable character of some experimenters. One of these was a joint author of the Handbook for the Physiological Laboratory. A very terrible experiment was quoted in that book; and being asked how he accounted for any humane person inserting it as an illustration, Professor Rolleston said that Dr. Foster had never shown it (the experiment on recurrent sensibility), and never seen it himself. Asked: "But surely it is put here, in a Handbook, in a mode which would encourage the trying of that experiment?" The reply was: "Obviously; but I am speaking in vindication of the character of my friend, but not at all in vindication of the book." Asked: "Then I understood that your opinion about the book is, that it is a dangerous book to society, and that it has warranted, to some extent, the feeling of anxiety in the public which its publication has created?" "I am sorry," replied Professor Rolleston, "to have to say that I do think that is so."

Others have shown no reserve at all in defence of everything contained in it, and have exhibited a defiance of public opinion too plainly arising from callous indifference. That they should be supported by men of more gentle and refined nature only proves the more strikingly that the tendency of vivisection is to blunt the moral sense. It is a law in ethics, that the strength of any motive is increased or diminished, according to the habitual exercise of the mental emotion brought into play. Sympathy for distress and aversion to inflict pain may be naturally strong in the heart of a biologist or physician, but may be gradually overpowered and suppressed by the habitual exercise of other motives, such as zeal for science or ambition of scientific fame. Every time these passions prevail an increased purchase is gained for their future influence, and the heart is hardened as they encroach on the rightful domain of sympathy and compassion for poor suffering animals. In other persons, the better feeling of possibly rendering good to men by improvements in medicine, represses the immediate emotion of pity; and even humane physicians advocate the most fearful proceedings of vivisection. Such is the natural process by which the feelings are blunted and the moral sense restrained from protesting against the cruelty of vivisection.

While thus explaining the personal blunting of feeling towards animals, in some who may be amiable and kind to their fellow-men, no reserve should be maintained in declaring the evil tendency of the system. To those who possess the large Blue Book, with the reports and evidence of the Royal Commission, or who have, in other ways, specially become acquainted with the history of vivisection, it would be needless to offer proofs on this matter. But a large proportion of the medical men of the day know little of what has passed in regard to the teaching of physiology in recent years. This is a new feature in English medical education. There were no physiological laboratories, not even class demonstrations, in our student days, at Guy's or St. Bartholomew's; nor at the Universities of Edinburgh or London was the practice of vivisection recognized. The altered attitude of the medical press, and of the official representatives of the profession, already show signs of deterioration of moral and social tone, and there is need for plainly showing the influences now at work, and leavening the character of the rising race of medical practitioners.

Dr. George Hoggan published in *Fraser's Magazine,* for April, 1875, a statement of what he had witnessed as assistant in the laboratory of one of the most eminent physiologists of France. The name is courteously withheld, but it is very well understood to what place the reference is made. " In that laboratory," says Dr. Hoggan, " we sacrificed daily from one to three dogs, besides rabbits and other animals, and after four months' experience, I am of opinion that not one of these experiments was justified or necessary. The idea of the good of humanity was simply out of the question, and would have been laughed at, the great aim being to keep up with, or get ahead of, one's contemporaries in science, even at the price of an incalculable amount of torture needlessly and iniquitously inflicted on the poor animals.

" During three campaigns I have witnessed many harsh sights, but I think the saddest sight I ever witnessed was when the dogs were brought up from the cellar to the laboratory for sacrifice. Instead of appearing pleased with the change from darkness to light, they seemed seized with horror as soon as they smelt the air of the place, divining, apparently, their approaching fate. They would make friendly advances to each of the three or four persons present, and as far as eyes, ears and tail could make a mute appeal for mercy eloquent, they tried it in vain. Even when roughly grasped and thrown down on the torture trough a low complaining whine at such treatment would be all the protest made, and they would continue to lick the hand which bound them till their mouths were fixed in the gag, and they could only flap their tail in the trough as their last means of exciting compassion. Often when convulsed by the pain of their torture this would be renewed, and they would be soothed instantly on receiving a few gentle pats. It was all the aid or comfort I could give them, and I gave it often. They seemed to take it as an earnest of fellow-feeling, that would cause their torture to come to an end—an end only brought by death.

" Were the feelings of experimental physiologists not blunted, they could not long continue the practice of vivisection. They are always ready to repudiate any implied want of tender feeling, but I must say that they seldom show much pity; on the contrary, in practice they frequently show the reverse. Hundreds of times I have seen, when an animal writhed with pain, and thereby deranged the tissues, during a delicate dissection, instead of being soothed it would receive a slap and an angry order to be quiet and to behave itself. At other times, when an animal had endured great pain for hours without struggling or giving more than an occasional low whine, instead of letting the poor mangled wretch loose to crawl painfully about the place in reserve for another day's torture, it would receive pity so far that it would be said to have behaved well enough to merit death ; and, as a reward, would be killed at once by breaking up the medulla with a needle, or 'pithing,' as this operation is called. I have heard the Professor say, when one side of an animal had been so mangled, and the tissues so obscured by clotted blood that it was difficult to find the part searched for, 'Why don't you begin on the other side ?' or, 'Why don't you take another dog ?' 'What is the use of being so economical ?'

" One of the most revolting features in the laboratory was the custom of giving an animal on which the professor had completed his experiment, and which had still some life left, to the assistants, to practice the finding of arteries, nerves, etc., in the living animal, or for performing what are called fundamental experiments upon it—in other words, repeating those which are recommended in the laboratory handbooks."

Such was Dr. Hoggan's experience in the laboratory of one who was in the first rank in Paris as a physiologist. His words are worth repeating. "I am of opinion that not one of those experiments on animals was justified or necessary." The wonder is how he could have assisted at such scenes of torture, as he calls them, for so long a period. It is well that he has now made so clear a statement and generous a confession. His evidence may serve as a warning as to what is possible in England, if this system of research spreads among us. Another English surgeon, visiting a French laboratory, describes the conduct of the students, in mimicking the cries and moans of the tortured animals in derision, as so revolting that he quitted the place in disgust. I myself witnessed, long ago, this "tiger-monkey" spirit in Magendie's classroom. Along with the late Edward Forbes, and two or three other students from Edinburgh, I tried to learn something from Magendie, but we were driven from the place in disgust, shocked, not so much by the coarse cruelty of the Professor as by the repulsive heartlessness of the spectators. English students were not in those days accustomed to such scenes of horror. The foreign teachers know the greater sensitiveness of our countrymen, although the honorable distinction seems to be passing away. An English student having quitted a well-known German laboratory, unable to bear its horrors, the professor said that "he never found Englishmen who would stop with him, and he supposed (with a sneer) that they thought God would make them suffer the same as the animals."

The experience of the last few years sadly proves how soon and how effectually the tone which has distinguished English from Continental schools has been lowered. Ten or twelve years ago Mr. Fleming, author of the first prize essay published by the Society for Prevention of Cruelty to Animals, after describing the fearful cruelties daily witnessed at the College of Alfort, the chief veterinary training school in France, could say that, "To the honor of the veterinary schools of England, vivisection has never been allowed in them;" and Mr. Fleming, with just pride, adds, "No one will deny that they are as well qualified to undertake the management of difficult operations as the vivisectionists." The

details of the practices at Alfort, and also at Lyons, as given by
Mr. Fleming, form a most ghastly record. The scandal caused by
these atrocities led to an appeal being made to the late Emperor of
the French, who referred the matter to a Scientific Commission.
The practices are, however, continued to the present day, and, we
grieve to say, have been introduced into this country. Mr. James
Mills has put on record a fearful account of cruelties which he witnessed, and in which he took part when attending the Edinburgh
Veterinary College, but of his share in which he is now heartily
ashamed. Both veterinary and medical students joined in the
experiments which Mr. Mills describes He says, " There was no
other motive than idle curiosity, and heedless, reckless love of
experimentation. To observe the heart's action a cat was fastened
down on its back. An incision through the skin of the animal's
chest extended from the neck to the belly. The skin was then
laid back by hooks, to enable the operator to cut through the cartil-
age of the sternum, and to draw his knife across the ribs for the
purpose of nicking them. The ribs were then snapped, and the
fractured parts turned back and secured by hooks. No anæsthetic
was used. On another occasion a horse was bought for the purpose
of dissection. During a whole week this animal was subjected to
various operations, such as tenotomy, neurotomy, etc., again without anæsthetics. In other cases the animals received " brutal
usage." Mr. Mills exonerates the professors from participation in
the experiments, most of which were performed in the students'
lodgings ; but the Principal must have known of the horse being
experimented on within the walls of the College. It is not surprising that Dr. Haughton, of Dublin, in his evidence before the
Commission, said : " I would shrink with horror from accustoming
large classes of young men to the sight of animals under vivisection.
I believe that many of them would become cruel and hardened, and
would go away and repeat those experiments recklessly. Science
would gain nothing, and the *world would have let loose upon it a set
of devils.*"

Dr. Acland, of Oxford, said, in his evidence, that many persons
are now engaged in the pursuit of vivisection in this country, not
for a humane purpose, but for acquiring abstract knowledge. This
desire of mere discovery has a dangerous and mischievous tendency.
" So many persons have got to deal with those wonderful and beau-

tiful organisms just as they deal with physical bodies that have no feeling and consciousness." Dr. Acland said this could not be done without being so hurtful to the moral sense of England that it would not be endured if carried to the same extent as abroad. Surely an effort must be made to prevent our English schools of medicine being degraded to the Continental level.

Much has been said about the evidence of Dr. Klein, Director of the Brown Institution, and Lecturer on Histology at the Medical School of St. Bartholomew's Hospital. He certainly made some candid and strange admissions as to the cruelties alleged to have taken place in his researches. He said that a physiologist could not be expected to devote time and thought to inquiring what the animal feels while he is doing the experiment. He "uses anæsthetics only for convenience sake, in dogs and cats, and for no other animals as a general rule." Dr. Klein must not be too severely judged. His training has been different from that of most English-men ; and he never knew in Vienna, where he formerly practiced, any of the hostility to vivisection which is common in this country on the part of the general public, though not of physiologists.

But Dr. Klein's statements lead us to view with dark foreboding the avowed opinions of some of our leading professors and public teachers, as when Dr. Burdon-Sanderson says he "wishes to see the type of education here more like the type of education in Germany." Dr. Gamgee, of Manchester, also praises highly the proceedings of Dr. Ludwig, of Leipsic, who has been the teacher of nearly all the physiologists of Europe, and has indoctrinated nearly the whole of them in the methods of physiological inquiry. These expressions of opinion, from prominent and representative men, and still more, the reported proceedings of the General Medical Council and of the British Medical Association, in reference to legislation on the subject, give rise to sad forebodings for the future. The new generations of medical men, trained under such influences, although few of them may have been personally engaged in experiments, must become degraded in moral and social tone, and the whole status of the profession will thereby be affected.

Foolish things may have been said, and extreme views held by those who advocate the total abolition, or suppression by law, of experiments on living animals. Even those who most wish this can scarcely hope to see their wish realized. But I do not despair

to see such a change in the general opinion of the profession regarding such experiments as will render them of rare and exceptional
occurrence. Apart from any ignorant clamor there is a strong
public feeling as to the cruelties of vivisection. Sir Arthur Helps
gave expression to the feeling prevalent among men of culture in all
professions, when he said that "any man known to have practiced
needless cruelties on animals should be placed under a social ban."
It is very certain that the status of the profession may be lowered
by being associated in the public mind with vivisection. There
are already signs of this, and many medical men would rejoice to
see their profession delivered from the opprobrium that has come
upon it in consequence of this practice. This can be done only by
showing that sound science is on the side of humanity on this
question. So far from vivisection having aided in the advancement
of the healing art, many testimonies confirm the saying of Sir
Charles Bell, that "it has done more to perpetuate error than to add
to sound knowledge." At all events the advantages of such experiments have been vastly overrated, and their disadvantages not duly
considered. The question is not whether any results are obtained
from this source, but whether they are worth the price paid for
them. That knowledge is dear which is purchased at the expense
of humanity. These experiments involve much suffering and
wrong, afford very meagre and doubtful results for practical use,
and withdraw attention from sounder methods of research. They
are neither scientifically valuable, nor morally justifiable.

www.ingramcontent.com/pod-product-compliance
Lightning Source LLC
Chambersburg PA
CBHW032347020726
47499CB00009B/3195